BUSINESS/SCIENCE/TECHNOLOGY DIVISION
CHICAGO PUBLIC LIBRARY
400 SOUTH STATE STREET
CHICAGO, IL 60605

QB
462.65
.H4613
1997

Chicago Public Library

25639265

Tower : an intertextur

D0855820

checkout service.
Thank you for using CPL's self
Please retain for your records.
Overdue: 0
Tuesday, August 09, 2016 6:33 PM
Account balance: $0.00
Total items: 1

Due: Tuesday, August 30, 2016
Id: R0125639265

Chicago Public Library

REFERENCE

Form 178 rev. 1-94

THE EINSTEIN TOWER

WRITING SCIENCE

EDITORS Timothy Lenoir and Hans Ulrich Gumbrecht

THE EINSTEIN TOWER

An Intertexture of Dynamic Construction, Relativity Theory, and Astronomy

Klaus Hentschel

TRANSLATED BY Ann M. Hentschel

STANFORD UNIVERSITY PRESS

STANFORD, CALIFORNIA 1997

Stanford University Press
Stanford, California
© 1997 by the Board of Trustees of the
Leland Stanford Junior University
Printed in the United States of America

*The Einstein Tower: An Intertexture of Dynamic Construction,
Relativity Theory, and Astronomy* is an expanded and revised
version of the original German *Der Einstein-Turm*
(Berlin: Spektrum Verlag). © 1992 by Klaus Hentschel.

CIP data are at the end of the book

BUSINESS/SCIENCE/TECHNOLOGY DIVISION
CHICAGO PUBLIC LIBRARY
400 SOUTH STATE STREET
CHICAGO, IL 60605

R0125639265

To my dear mother,

Ruth Schmidt-Stockhausen

BUSINESS/SCIENCE/TECHNOLOGY DIVISION
CHICAGO PUBLIC LIBRARY
400 SOUTH STATE STREET
CHICAGO, IL 60605

Acknowledgments

Special thanks go to Prof. Hermann A. Brück, of Edinburgh, for permission to publish two photographs in his possession, and to Mr. Winfried F. Freundlich, of Wiesbaden, for personal reminiscences about his uncle. My thanks for constructive comments and criticism concerning topics related to this book go to Dr. Peter Ackermann, Prof. Lorraine Daston, Dr. Wolfgang R. Dick, Prof. Bernhard Hassenstein, Hans G. Heck, Jost Lemmerich, Mr. Wolfgang Lille, Prof. Wolfgang Mattig, Prof. Christoph Meinel, Dr. Karl von Meyenn, Prof. Dr. H. Montag, Prof. John D. North, Dr. Skuli Sigurdsson, Dr. Jürgen Staude, and Prof. C. A. Truesdell. I also appreciated John W. Briggs's very impressive demonstration of the coelostat system of the Snow telescope at the Mt. Wilson Observatory and the kind invitation of Ms. Barbara Eggers and Prof. Karl-Heinz Rädler to take part in the celebration of the seventieth anniversary of the inauguration of the Einstein Tower in Potsdam in 1994.

Permissions to cite unpublished documents were generously granted by: Ms. Schaumberg of the Mary Lea Shane Archives at the Lick Observatory, Mt. Hamilton, California; Dr. Henning at the Archive of the Max Planck Society, Berlin-Dahlem; Dr. Knobloch at the Central Archive of the former East German Academy of Sciences; Dr. Glaus of the Library of the Eidgenössische Technische Hochschule, Zurich; Mr. Wagner of the German Museum of Architecture, Frankfurt am Main; Mr. Heverly of the Archive for Scientific Philosophy in the Twentieth Century, Pittsburgh; Dr. Rohlfing of the Manuscripts Department of the State and University Library, Göttingen; Dr. Hunger of the University Archive, Göttingen; Prof. Evers of the Fine Arts Library of the Staatliche Museen Preußischer Kulturbesitz, Berlin; Mr. Rosenkranz of the Albert Einstein

Archive, Hebrew University of Jerusalem; and Ms. Edwards on behalf of the Royal Society, London. Permissions for photographs and illustrations were also obtained from the Bettmann Archive, New York; Ullstein Bilderdienst, Berlin; Ernst Wasmuth Verlag, Tübingen; and from Springer Verlag, Heidelberg.

I would also like to thank Mr. von Knobelsdorff of the Berlin Technical University, Dr. Schultze of Humboldt University, Berlin, and Dr. Siggemann of Johannes Gutenberg University, Mainz, for their research efforts in tracing Freundlich documents at their associated university archives. Many thanks also to the many archivists who assisted me during my research stays or who prepared and mailed requested photocopies.

Finally, I would like to thank Prof. Timothy Lenoir, the editor of the series in which this volume appears, for his support, Helen Tartar, Humanities Editor, Amy Klatzkin and Nathan MacBrien, Associate Editors, at Stanford University Press, and Paul Psoinos, as copy editor, for their invaluable assistance in the final preparation of the manuscript.

Contents

✦✦✦

Reference Matter

Figures

Tables

THE EINSTEIN TOWER

Introduction

The three main themes of this book may appear to be equally obscure: the theory of relativity has had, among nonspecialists, the unmerited reputation since its conception of being completely incomprehensible; the first German astronomer to explore the consequences of Einstein's theories for his field, Erwin F. Freundlich, is also familiar only to specialists today; finally, few will have had occasion to view both the interior and exterior of the Einstein Tower, an architecturally historic observatory on the Telegraphenberg ("Telegraph Hill") in Potsdam. Freundlich was the first German astronomer to take an interest in the experimental verification of the general theories of relativity and gravitation. However, like the theories themselves, he became a highly controversial figure and he could only pursue this goal through Albert Einstein's personal intervention. Freundlich became the first full-time staff scientist at the Kaiser Wilhelm Institute of Physical Research. To afford him more independence, the decision was finally reached in 1920 to establish a separate institute for the astronomer, where he could concentrate on examining Einstein's postulated effect of gravitational redshift in the solar spectrum. This institute consisted essentially of an observatory and attached laboratory and soon came to be known informally as the Einstein Tower. Freundlich managed to engage the interest of his friend Erich Mendelsohn to design this unique tower-shaped building, which eventually brought the young architect international fame.

There will be many layers to separate out of the narrative, which will prove indispensable to understanding the whole picture. Any attempt to lessen the significance of any one of them would lead only to an unbalanced depiction of the historical events. The various interwoven descriptive levels forming the fabric of this book include the following biographical and cognitive elements and institutional and disciplinary histories:

The protagonist's biography (Freundlich's origin, religion, education, employment);

The social context, covering his interaction with academic teachers (particularly Felix Klein), coworkers (H. von Klüber, A. von Brunn, H. A. Brück, E. von der Pahlen, W. Grotrian, A. Unsöld), and students (e.g., A. H. Batten), and his patrons (A. Einstein, M. Planck) and opponents in the scientific community (H. Struve, H. von Seeliger, H. Ludendorff);

The cognitive aspect, including a survey of his scientific papers and focusing on his attempts to verify the general theory of relativity through observation;

The political milieu within the Berlin scientific research community;

A transnational comparison of astrophysics, particularly noting American observatories;

The Einstein Tower telescope within the context of architectural history (technical specifications, architectonic layout and design, stylistic designation and comparison with other structures of the time), together with

Economic and sociopsychological components (motives behind and arguments for its construction; financing issues linked with the interwar inflationary economy);

The reception of the Einstein Tower and relativity theory—portentous symbols of modernity or bizarre outgrowths of a "degenerate" culture?

A historical examination of the research results generated at the Einstein Tower during the Freundlich era, also set against the tower's original purpose;

The rise of National Socialism and its impact on the Einstein Institute: Ludendorff's revenge and Freundlich's exile and subsequent fate.

We must start with one of the above descriptive levels (in this case, the biography of the figure who appears to be the protagonist) and then extend outwards into other levels. A glance at the opposing groups of benefactors and enemies of Freundlich will reveal the special role they played in the Berlin scientific community. It will be shown that the construction of the Einstein Tower created a research niche for the professionally and politically isolated E. F. Freundlich, whose aggressive advocacy of empirically testing Einstein's general theory of relativity caused him to lose his footing completely among the predominantly conservative German astronomers and astrophysicists.

Added perspective will be gained from establishing the social and institutional context of Freundlich's career during his Berlin and later years. Thus Freundlich serves as a kind of test particle: the numerous turns along the course of his career reveal the attractive and repulsive potentials of the Berlin research environment. In contrast, Einstein's "total mass" was too ponderous—that is, he was too famous, too idiosyncratic, and otherwise too well protected—to be swayed significantly up to 1933. Freundlich's winding path, on the other hand, fully reflects the impact of the various powerful scientist cliques.

It took the political upheavals of 1933 throughout Germany to uproot both the giant in science politics Einstein and the dwarf Freundlich, along with thousands of other intellectuals who could no longer see any future in Germany (among these the Einstein Tower's architect, Erich Mendelsohn). The National Socialist seizure of power brought about a change in the research program conducted at the promptly renamed "Institute of Solar Physics." Chapters 10 and 11 reveal that this abrupt reorientation at Potsdam on the outskirts of Berlin had its roots in the personal enmity between Freundlich and Ludendorff, which dated back to well within the Weimar period. In 1933 the National Socialist Ludendorff took over the helm. After having had to bend to Einstein's and von Laue's influence in decisions made at the board meetings of the Einstein Foundation (Einstein-Stiftung) and the Astrophysical Observatory (Astrophysikalisches Observatorium), Ludendorff was now finally able to enjoy to the fullest being himself in the position of power. The international contacts Freundlich had laboriously established (see Chap. 6) were broken off and his research agenda (see Chap. 8) overturned. After 1939 the research was even tuned to the needs of the military: regular solar observations were made to enable the prediction of electromagnetic transmission disturbances, which often follow powerful outbursts on the surface of the sun.[1]

For a well-balanced description of the events of 1933 we must take a look at not only the general political and economic conditions but also the social psychology of the time, just before the construction phase of the Einstein Tower. It is astonishing that in the disruptive interwar period private donors and even the state were willing to part with the significant sums necessary for the construction and maintenance of an institute with such esoteric goals. The explanation lies in a unique combination of factors that only came together in 1919:[2]

The prevalent fear among German scientists of losing ground to foreign research (particularly in the United States), reinforced by

The shocked realization that a British solar eclipse expedition in 1919[3] had verified a light deflection hypothesis (see Chap. 1) by Einstein, one of their own scientists;

Considerable public interest in Einstein's theory of relativity, fanned by the correspondingly extensive coverage of the light-deflection measurements in the new mass media;[4] and finally

An increase in private donations for the advancement of science since the beginning of this century, primarily for publicity reasons (see Fig. 33). The Kaiser Wilhelm Institutes in particular received considerable contributions in support of research otherwise conducted at state-funded universities.[5]

Freundlich's biography forms the backbone of this book. However, the characteristics of the *Denkkollektive* (in Ludwik Fleck's sense) in which Freundlich's activities were embedded or the science communities (as defined by Joseph Ben-David or Thomas S. Kuhn) against which these activities were directed come repeatedly to the fore. A new fabric should emerge from these many singular strands combining biographical, social, institutional, and cognitive elements to provide at least an outline of the thematic complex "Einstein Tower, Freundlich, and relativity theory." Although I have confined myself to the ten levels listed above, I am aware that other layers ought to be included in a truly comprehensive description. The interaction between the Einstein Tower as a scientific institution and a major optical manufacturer (Carl Zeiss in Jena) during the planning and installation stages of the tower had to be omitted, for example, for lack of documentation either in Jena or in Oberkochen.[6] Erich Mendelsohn's place in architectural history was only covered briefly, considering that in-depth studies on Mendelsohn's work already exist or are in preparation.[7] I was able to examine the correspondence between Einstein and Freundlich, though, which was auctioned a while ago and copies of which are also on file at the archive of the Collected Papers of Albert Einstein in Boston.[8] Other letters by Freundlich were located in the estates of various scientists with whom the astronomer had been in contact. I was unable to discover the whereabouts of Freundlich's diaries, his private correspondence, and the like. Nonetheless there is an abundance of documents by and on Freundlich upon which this study is based. A list of the archives that I consulted precedes the endnotes to this volume.

CHAPTER 1

The Young Astronomer
Erwin Finlay Freundlich

Six years Einstein's junior, Erwin Finlay Freundlich (1885–1964)[1] was born in Biebrich on the Rhine, the son of a German manufacturer, Friedrich Philipp Ernst Freundlich, and Ellen Finlayson, from Cheltenham, England. Freundlich's grandmother was a practicing Jew; but her son converted to the Protestant faith at the time of his marriage. His seven children all were baptized and had a Protestant upbringing. The oldest son was the noted colloid chemist Herbert Max Freundlich (1880–1941).[2] His younger brother Erwin had a normal education: after attending technical secondary school (*Realgymnasium*) to the third year (*Quarta*), he received his diploma (*Abitur*) at another high school in the humanities, graduating at Easter in 1903. A half-year of voluntary service at the Vulkan shipping wharf in Stettin inspired the young man to study shipbuilding at the Charlottenburg Polytechnic in Berlin. He abandoned this subject after two terms, however, in the autumn of 1905 in order to study mathematics and astronomy at Göttingen (8 terms),[3] transferring briefly to Leipzig in his third term. The astrophysicist Karl Schwarzschild (1873–1916) and the famous mathematician and science administrator Felix Klein (1849–1925) were among his teachers.[4] Freundlich chose a topic on function analysis for his doctoral thesis on the suggestion of a function theory specialist, Paul Koebe (1882–1945). The thesis, entitled "Analytic Functions with Arbitrarily Set Domains of Infinitely Many Sheets" (Analytische Funktionen mit beliebig vorgeschriebenem, unendlichblättrigem Existenzbereiche), was submitted on January 26, 1910; and Klein gave it the good but not outstanding mark III (*opus laudabile*).

Klein's grade for the oral mathematics examination was even lower (*rite*). In explaining this merely satisfactory grade Klein remarked how difficult it was "to get really precise answers out of him"; but he also noted Freundlich's "lively interest in many areas of mathematics." Freundlich's examiner for physics was Woldemar Voigt, and for astronomy Johannes Hartmann.[5]

When at the end of July Klein procured the graduate a position at the Royal Observatory (Königliche Sternwarte) in Berlin as assistant to the director, Karl Hermann Struve (1854–1920),[6] Freundlich reportedly protested to his teacher that he did not know enough about practical astronomy. Klein's categorical reply was: "You did not come to university in order to learn everything but in order to learn how to learn everything. You are to go to Berlin."[7] This is in line with Felix Klein's active promotion of closer ties between mathematics and the physical and engineering sciences. Both he and a number of his students made important contributions to these disciplines and their peripheral fields.[8] An argument for recommending his mathematically trained student to the Berlin Royal Observatory is also provided in David Hilbert's famous statement "Physics is too important to be left to physicists" when applied to the field of astronomy. Nevertheless, for Freundlich the move from Göttingen to the Prussian capital must have been like a plunge into cold water. He was immediately immersed in routine duties at the observatory (e.g., the compilation of a zone catalog of pole stars, photometric observations, as well as work using a meridian circle), which were not particularly intellectually challenging for the young man.[9] A welcome change came in August 1911, when the Prague astronomer Leo Wenzel Pollak (1888–?)[10] made inquiries on behalf of the then newly appointed young professor of theoretical physics at the German University in Prague, Albert Einstein. Freundlich's attention was thus turned to the question of how Einstein's new theory, introduced in his paper "On the Influence of Gravitation on the Propagation of Light,"[11] could be tested using astronomical means.[12] As Freundlich learned from Pollak, Einstein was of the opinion that two of the postulated relativistic effects, gravitational redshift and light deflection, "urgently [needed] verification through astronomy." Einstein also pointed out that such a verification could possibly succeed only from examining the sun, because of the dependence of these effects on the magnitude of the gravitation potential, $\phi \sim m/R$.[13]

Because Einstein's theory of space, time, and gravitation leads to physical conclusions different from those of classical physics, which was then

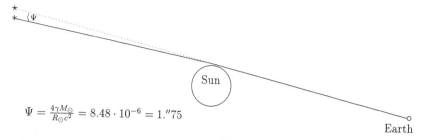

$$\Psi = \tfrac{4\gamma M_\odot}{R_\odot c^2} = 8.48 \cdot 10^{-6} = 1.''75$$

Fig. 1. The deflection of light in the sun's gravitational field. Light rays passing close by the sun are deflected, thus making their source appear shifted away from the solar rim. The slight shift in the apparent position (★) of a star in comparison with its actual position (*) is indicated by the angle Ψ. (This drawing greatly exaggerates the effect.)

generally accepted, proving the existence of the two physical effects above would be a significant step.[14] The theory postulates that the presence of large masses has a particular influence on the propagation of light, revealing a curved space-time structure.[15] Einstein predicted the size of the effect in relation to the size of a mass and its proximity to passing light. With the means then available, the sun, the central body of our planetary system, was the largest mass with which this could conceivably be observed. It was hoped that the deflection could be observed in light originating from remote fixed stars as it passed by the rim of the sun. This idea was not so unfamiliar. In 1906 Einstein had already proposed that energy and mass are physically equivalent,[16] and it had been known since the end of the nineteenth century that light transports energy E. In 1905, Einstein demonstrated that energy E is proportional to its frequency v. Thus it should also behave like a current of particles of mass $m = E/c^2$ (with c as the velocity of light). Therefore, stellar rays of light should be slightly deflected from their course as they pass near the surface of the sun, just like passing comets. After 1913 Einstein modified this concept of light deflection to conform it to the altered space-time structure near large masses, or curved space. His final calculation of the deflection values was disappointingly small, though. According to Einstein's 1911 theory, the deflection of rays passing the sun tangentially measured $0''.85$ degrees, that is, less than one arc second. The value offered by the 1915–16 theory was $1''.75$ degrees.[17] As distance r increases, the effect decreases by $1/r$. (See Fig. 1.)

The presence of mass influences not only the structure of space but time measurement as well, according to Einstein. Clocks and periodic

Fig. 2. The new Royal Observatory building at Babelsberg, near Berlin, completed in 1913. (From Struve 1919)

processes in general slow down when they are brought in proximity to a mass. Gaseous atoms exhibiting a specific and exactly determined eigenfrequency can thus serve as clocks that perform not sixty but millions of oscillations per minute. This periodicity is detectable in the light they emit or absorb, appearing as sharp lines in the spectrum. Analyses of the solar spectrum have shown that many elements found on earth also exist in the sun; accordingly, Einstein hoped that a precise comparison of the solar spectra of these elements against laboratory spectra would reveal an apparent shift of the absorption lines in the solar spectra toward the lower frequencies as against their terrestrial counterparts (i.e., toward the red end of the spectrum) as a direct result of the sun's mass. Yet this effect, known as gravitational redshift, is also minute: detecting relative shifts of the spectral lines on the order of less than one hundred-thousandth $(2 \cdot 10^{-6})$ was at the very limit of precision that spectroscopic measurements of the time could obtain. Nevertheless, Einstein did find in the scientific literature a few hopeful indications of the existence of such a shift in the solar spectrum.[18] Since the relativistic effects were so minuscule, it is no wonder that astronomers practically ignored Einstein at first. There was one exception in Germany: in contrast to other astronomers in Berlin and elsewhere, Erwin F. Freundlich was immediately drawn to Einstein's theories, clearly as a result of his more up-to-date mathematical training at Göttingen.[19] But as Freundlich himself writes in a résumé around 1915, he was only able to work on his "analyses of a theoretical nature," specifically, "to test modern physical theories by astrophysical means," in his "hours of leisure."[20] Nonetheless, in 1913 the focus of his research efforts began to shift to the various more inspiring astronomical consequences of Einstein's theories.

Fig. 3. Meridian circle at the Berlin Royal Observatory used by Freundlich after 1910. (From Struve 1919)

Fig. 4. Two portraits of Freundlich: *left*, about 1937; *right*, about 1960. (From *International Portrait Catalogue of the Archenhold Observatory* [Berlin: Archenhold Obs., 1984], by permission of Prof. Dieter B. Herrmann)

How could one of Struve's assistants succeed in seriously neglecting the duties for which he had been employed at the Berlin Royal Observatory, to become so intensely involved with minute effects predicted by a theory that all astronomers in Berlin and elsewhere disregarded? A quite frank letter to William Wallace Campbell, director of the Lick Observatory on Mt. Hamilton in California, provides one answer. After briefly describing the effect of light deflection in a gravitational field, Freundlich reports in his imperfect English:

Mr. Einstein asked me, if I would try to proof his results by observations. I naturally assented as I have been studying these questions allready for a long time.

At the same time observations of Mr. Courvoisier at Berlin seem to guaranty an influence of the sun-distance upon the position of a star in Rightascension and declination; but this effect is not yet quite explainable and sur[e]ly of a different nature than the effect deduced by Mr. Einstein. Therefore I have began to investigate these rather important questions.[21]

Einstein's great influence on Freundlich is documented in the correspondence initiated between the two in 1911. At this time Freundlich was conducting investigations on star positions with Leo Courvoisier (1873–

1955),[22] who brought to his attention apparent fluctuations in the position of fixed stars.[23] Since 1905 Courvoisier had pointed out repeatedly that the positions of fixed stars are subject to a small but systematic deviation when located close to the sun's disk (as seen from earth), which effect he named "annual refraction."[24] In connection with his subordinate menial duties at the Berlin Royal Observatory, Freundlich was assigned the task of completing and evaluating Courvoisier's data in 1912. Although Courvoisier and Einstein had arrived at different predictions for the apparent position shifts of distant stars, there was nevertheless a certain qualitative similarity between the two hypothesized effects. Both decreased with apparent distance from the sun. It is noteworthy that Courvoisier attributed annual refraction to the refraction of light passing through a denser medium surrounding the sun—the ether—in conformance with classical electrodynamics. From his data Courvoisier derived the earth's speed in motionless ether at around 650 km/sec.[25] But by 1905 Einstein had already rejected the ether hypothesis as "superfluous" in his paper on the electrodynamics of moving bodies. Freundlich found it much more intriguing to establish which of these effects actually existed, thus basically pitting Courvoisier's traditional physics, incorporating ether, against Einstein's "modern" theory of relativity. Numerous shaky predictions in Courvoisier's theory about the circumsolar medium damaged its plausibility. These included the decrease in the medium's density with increased distance from the sun (and thus decrease in refraction), the medium's resistance to motion, and its absorptive properties, none of which agreed with existing observational data.[26] Thus Freundlich had a strong incentive to become immersed in Einstein's theory and its experimental consequences. This theory could eliminate the need for the questionable results attached to Courvoisier's own interpretation of annual refraction.

CHAPTER 2

A Theorist's Observer: Freundlich's Collaboration with Einstein from 1911

Freundlich must have responded immediately to Pollak's initial inquiry, almost as if he had been waiting for such a diversion to liberate him from the endless columns of numbers in his meridian observations. But Einstein was equally delighted about Freundlich's prompt reaction. Addressing the lowly assistant in Berlin as "Esteemed Colleague," the Prague *Ordinarius* wrote: "Thank you very much for your letter, which naturally interested me greatly. I would be personally very pleased if you took up this interesting question."[1] Freundlich proposed the idea of inspecting existing solar eclipse photographs and the corresponding background sky images for the apparent small position shifts of fixed stars Einstein had predicted.[2] In response Einstein reiterated his continued interest in the results of this test: "I am extremely pleased that you have taken on the question of light deflection with such enthusiasm and am very curious what the examination of the available plates will yield. This question is of very fundamental importance. From the theoretical standpoint it is quite probable that the effect really exists."[3]

The "fundamental importance" of verifying light deflection, or whether light is subject to gravitation, was that Einstein's special theory of relativity of 1905 implies the equivalency of energy and mass ($E = mc^2$). Since light transports energy, it also should exhibit the properties of inertia, just as do masses in a gravitational field. Einstein's paper of 1911 gave the first systematically derived estimate for the effect of gravitational fields on the propagation of light emitted by fixed stars passing close by the sun's rim. (See Chap. 1 and Fig. 1.) This prediction could be verified directly during an eclipse by measuring the exact position of fixed stars closest to the darkened solar disk, which are ordinarily hidden by the sun's glare. But solar eclipses are a relatively rare occurrence, and more often than not, to observe them under favorable conditions, special, extravagant, and costly expeditions must be undertaken to remote regions of the globe. This explains Freundlich's initial idea of consulting data from previous solar eclipses. When it became apparent that all the Potsdam plates were completely inadequate for his purpose (see below), Freundlich wrote to the major astronomical centers throughout the world, asking for quality glass-plate reproductions of earlier solar eclipse photographic exposures. His request to William Wallace Campbell (1862–1938)[4] of the Lick Observatory, dated November 25, 1911, has been preserved:

I apply to you on account of a question of sur[e]ly high scientific interest, in which I depend from the kind support of astronomers, who posses[s] eclipse-plates, if I shall hope to get any results. The modern theory of relativity of Mr. Einstein predicts an influence of any field of gravitation upon light passing near to the sun. The gravitation would have according to the investigation of Einstein the effect of deflecting the ray of the star, and Mr. Einstein asked me, if I would try to proof his results by observations. . . .

Now eclipse-plates of the sun showing at the same time a few stars are the only material to be used till now, and I would be very much obliged to you for your kind support to get such plates. A few days ago I had the opportunity to talk to Mr. Perrine at Cordoba on his way through Berlin and he gave me the advice to apply to you on account of this question.[5]

After many weeks of waiting Freundlich had still not received a response and directed another inquiry to California. At last on March 13, 1912, Campbell answered, regretting the loss of Freundlich's first letter in the post and assuring him of his ready assistance in his research project.[6] High-quality glass-plate reproductions were actually sent to him of exposures made during the solar eclipses of 1901 in Sumatra, 1905 in Spain, and 1908 on Flint Island, along with comparison photos of the stellar background. The most recent solar eclipse, in particular, had occurred against a backdrop of many hundreds of fixed stars. All plates made in these earlier expeditions had four serious drawbacks, however:

1. The solar image was always at the edge of the plate, not in the center.

2. The exposure time had been set for the brilliant solar corona, not for the distant and dimmer fixed stars.

3. Aiming adjustments during exposure were made to accommodate the sun's motion rather than the rate of the fixed stars in the background.[7]

4. It was impossible to investigate any contractions that the photo-plate might have experienced during exposure (determined, for example, by a sudden drop in temperature).[8]

Freundlich soon had to give up hope of being able to make use of existing plate material in his search for evidence of light deflection in the solar gravitational field. Camera lens distortions at the edge of the exposures as well as blurred star positions as a result of the camera's having been trained at an inappropriate speed rendered them completely unsuitable.[9] Freundlich reported his negative results in the journal *Astronomische Nachrichten* in 1913:

With the aid of an instrument by Toepfer (Potsdam), which Prof. K. Schwarzschild kindly made available to me, I began first of all to measure a plate from the Smithsonian Institution. . . . I measured the Cartesian coordinates of all the stars on the plate near the sun. But since the plate was neither sufficiently centered nor sharply focused and the stellar images were as a result very blurred, there turned out to be such great discrepancies in the settings for the stellar images on the different days that I had to give up using this plate. But it soon became apparent with all the exposures, even the very valuable ones from the Lick Observatory, which had all simply been produced for completely different purposes, primarily for the discovery of intramercurial planets, that the inadequate sharpness of the stellar images made any successful measurement of the plates illusory. . . . I have therefore delayed investigating the whole question until really usable material is available.[10]

When Einstein heard the bad news, he wrote to Freundlich: "Thank you very much for your detailed report and for your unusually keen interest in our problem. It is a great shame that the photographs now available are not sharp enough for such measurement."[11] Even though Freundlich could not produce a verification of the theory virtually overnight, as Einstein had perhaps originally hoped, he nevertheless did perform an important service. Through his articles in professional journals such as the *Astronomische Nachrichten* and *Physikalische Zeitschrift*, Freundlich announced the astronomical relevance of Einstein's theory to the German community of astronomers.[12] He was well in advance of de Sitter, who performed the same role for the English-speaking readers of the *Monthly*

Notices of the Royal Astronomical Society with his synopsis paper of 1917. Einstein recognized how indispensable this publicity was toward the general acceptance of his theory. Moreover, Freundlich's first publication in German solicited the collaboration of his colleagues ín testing Einstein's predictions. In a letter to his "highly esteemed colleague," at that time the lone relativity advocate within the ranks of professional astronomers in Germany, Einstein writes: "Thank you very much for your interesting letter. It will be thanks to your diligence [*Eifer*] if the important issue of the deflection of rays of light now awakens an interest among astronomers as well."[13]

But Freundlich had another idea on how to test the existence of light deflection: daytime observation of fixed stars in the background sky of the sun or Jupiter, thus not having to rely on the relatively rare occurrence of solar eclipses.[14] Although this second plan also fell through, Einstein was obviously delighted to have found someone with whom he could discuss at length the astronomical verifiability of his theoretical speculations. Another paragraph in Einstein's first letter to Freundlich illustrates this. There he refers to the second important consequence of his theory, the redshift of spectral lines through the influence of the sun's gravitational field, which according to his theory would only be on an order of magnitude of 0.01 Å (10^{-12} m).[15] But there were other possible explanations for this minor shift or asymmetric broadening of spectral lines:[16]

Unfortunately the broadening on both flanks of the spectral lines [graphs] depends on various causes (pressure—light dispersion [Julius]—motion [Doppler]), so that a convincing interpretation is hardly attainable. Do not very distinct solar lines also exist (that is, not over 0.02 Å thick)? I write you this in passing, incidentally, because I do not believe that clear results can be arrived at along this route.

Asking you to keep me informed about your investigations, for which I wish you luck, . . .[17]

It is striking that Einstein's estimation of the obstacles should have held true for so many years, even decades, although aside from Freundlich, other experienced observers, including John Evershed, Willem Henri Julius, and Charles Edward St. John at such various locations as Potsdam, Madras, Utrecht, and Mt. Wilson, competed to come up with clear experimental results on gravitation redshift in the solar spectrum.[18]

Freundlich adopted Einstein's rather skeptical view about the feasibility of testing the second of the two effects arising out of the Prague theory (see Chap. 3), and concentrated initially on light deflection. The

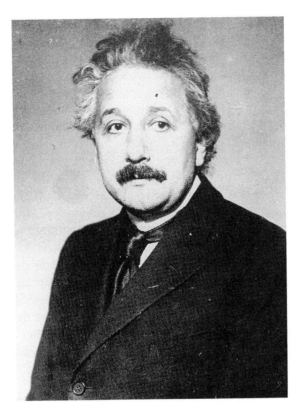

Fig. 5. Albert Einstein around 1914. (By permission of the Bettmann Archive, New York)

important by-product of Freundlich's efforts was his contact with other astronomers. Thus he became Einstein's mouthpiece in the world of astronomy. Thanks to Freundlich's inquiries, around 1913 the debate on the reception of Einstein's theory of relativity spread among astronomers at both the national and the international level. This is in stark contrast to other new theories of this period, which were discussed and developed within a local context for a much longer period of time before ever entering the global arena. Courvoisier's theory of annual refraction, for example, was hardly discussed outside the Astrophysical Observatory at Potsdam.

A theory cannot gain a foothold solely through the creator's efforts to introduce its strengths to the members of a scientific community to which he himself does not belong.[19] Through Freundlich's commitment to test relativity theory, Einstein was able to attract the attention of astronomers to his physical theory, revealing its relevance to precision measurement within their field. This explains Einstein's tenacity in encouraging Freundlich to seek astronomical tests of the general theory of relativity, even after the relations between the two men had cooled noticeably. (See Chap. 8.) The theorist's often daunting task of recruiting willing experimenters to test new predictions was a complete success in this case. Freundlich's own papers, in particular a semipopular overview article for the journal *Naturwissenschaften* in 1916, which was also published in that year by Julius Springer (and later translated into English and French), show that Freundlich had become completely conversant with Einstein's gravitation theory by this time:

The experimental foundation of Einstein's gravitation theory has thus not progressed very far yet. If the theory can nevertheless already lay claim to general acceptance today, this has its justifiable reasons in the extraordinary unity and logic of its basic tenets. With one stroke it truly solves all the puzzles that have arisen since Newton's time . . . in the description of natural processes from the motion of bodies.[20]

CHAPTER 3

Relativity Theory Under Scrutiny: Experimental Testing

Since 1907 Einstein had been working on expanding his 1905 special theory, and both theories were developing quite rapidly. The relativity principle of 1905 described only uniform linear motion, and the search for a more general theory of relativity to include acceleration and, as Einstein soon realized, gravitation as well lasted until November 1915.[1] In 1907 Einstein concluded that only with a strictly local equivalency between force and acceleration fields can accelerated forms of motion be included in a generalized theory. This arises from the idea that it is impossible to distinguish (from within a closed box or elevator, for example) between a downward pull caused by gravity and a numerically equal acceleration in the opposite direction (such as in a rocket at lift-off).

All Einstein's letters to Freundlich cited above refer to the second phase of development in the generalized theory known as the "Prague theory" of 1911. It examined two effects in particular: the gravitational redshift of spectral lines, and light deflection in the sun's gravitational field. (See Chap. 1.) In extending relativity to rotating reference systems after 1911, Einstein was forced to abandon the assumption that space was Euclidean. Tensor calculus was thus called for, and Einstein himself had first to be tutored in this mathematical tool by his former college classmate Marcel Grossmann. In a letter predating the first version of this tensorial theory of relativity and gravitation, later referred to as the "draft" theory, which he presented together with Grossmann in 1913,[2] Einstein reported to Freundlich: "My theoretical efforts are now making smart progress after an indescribably laborious search, so that all indications are that the equations on the general dynamics of gravitation will be drawn up soon. The nice thing about it is that it can be kept free of arbitrary assumptions, leaving nothing to be 'patched up'; the whole thing will be either true or false."[3] But it was only at the end of 1915 that the final general gravitation equations, or field equations, connecting the metrics of space and time with the distribution of mass and energy were discovered independently by Einstein and the Göttingen mathematician David Hilbert (1862–1943). Thus in retrospect the "indications" Einstein mentioned in his letter to Freundlich had been quite deceptive. The second sentence in the quotation above certainly is characteristic of Einstein's methodological approach, however. He attempted to construe the desired theory out of a few indispensable and plausible constraints. These included fulfillment of the conservation laws, absence of mathematical contradiction, approximate agreement with the classical formulas, and satisfaction of Mach's principle. (The latter, later abandoned, demands accounting for inertia in terms of the large-scale distribution of masses.) The logical interconnection of a number of these in themselves quite general conditions was restrictive enough to lead the theory to quite specific predictions: the magnitude of the spectral line shift and of the light deflection in the sun's gravitational field. Both these effects depended on the mass-to-radius ratio of a stellar body, and this was known quite accurately for the sun. Thus the predictions from Einstein's theories of gravitation could not be modified further, at least when applied to the sun. If any discrepancy with observation arose, no conceivably adjustable variables remained to salvage the theory. Its predictions were an all-or-nothing gamble: either the effects verifying his theory existed unequivocally, or else it was fundamentally flawed. He writes in this vein on gravitational light deflection: "But one thing can definitely be said: if such a deflection does not exist, then the theory's assumptions do not apply. It should be kept in mind that these assumptions, though likely, are certainly quite bold."[4] The special theory of relativity of 1905 was a "theory of principle" (*Prinziptheorie*), as Einstein called it, which—as distinguished from "constructive" theories, such as statistical mechanics—are based on "likely" but nonetheless "bold" axioms. Neither in his letters nor in his published papers did Einstein make any secret of the fact that these theories stood or fell with the principles on which they were axiomatically based. It was so important to him to forge ahead with the experimental consequences of these principles for precisely this reason; and he was quite prepared to give up his theory entirely if it were to prove contradictory to observation.[5] Einstein's correspondence with Freundlich iterates this. On expressing his curiosity about the results from Freundlich's binary star investigations, which tested the special relativity axiom of the independence of the velocity of light from the velocity of its emitter in a vacuum, he stated: "If the velocity of light is dependent even to the smallest extent on the velocity of the light source, then my entire relativity theory, including the theory of gravitation, is false."[6] This is surprising when one considers how easily Einstein could explain away experimental results in other situations or declare them insignificant when they did not suit him. Einstein attributed to both effects the same crucial importance because gravitational redshift was directly dependent on his equivalency principle, as was the light deflection hypothesis on the mass-energy equivalency. If these theoretical principles in which Einstein had such confidence were proved to be right, then the inferred effects must necessarily exist. "Theoretically the matter has come to a close of a sort.[7] I am secretly quite firmly convinced that light rays actually do undergo deflection."[8]

Einstein has thus already come to a decision in August 1913, having gone as far as thought experiments and an assessment of theoretical principles could take him. The only way remaining to find out whether his intuition was correct physically was through empirical analysis of the few detectable effects arising out of his theory. These subtle deviations in the behavior of test particles and light in the vicinity of larger masses would decide whether he would have to abandon his cherished principles and look for new ones, or whether he could continue to build upon them.

This image of Einstein the anxious hunter of empirical confirmation for his theories is quite in contrast to the one circulated by Ilse Rosenthal-Schneider, not coincidentally a Neo-Kantian. She portrays Einstein as dismissively indifferent to news in 1919 that a British expedition had verified his light deflection hypothesis. In her version he professed to have known all along that it would be confirmed eventually; and moreover that had it not been confirmed, only the experimenters' image would have suffered. In contradiction to this myth Einstein's correspondence not only with Freundlich but also with the Bonn physicists Leonhard Grebe and Albert Bachem,[9] for example, reveals downright eagerness for empirical proof at this stage.[10]

Even apart from Einstein's intuitive trust in a select number of fundamental physical principles, his confidence could only have been strengthened by a comparison of alternative contemporary theories explored primarily by Max Abraham (1875–1922),[11] Gunnar Nordström (1881–1923),[12] and Gustav Mie (1868–1957):[13]

My view of the other current theories of gravitation is the following. Abraham's theory, according to which light is deflected just as in mine, is inconsistent from the standpoint of invariant theory. The relativity theories of gravitation by Mie and Nordström remain. The former is fantastic, and in my opinion its intrinsic likelihood is infinitely small. But the latter is very reasonable and indicates a way, free from inconsistencies, of pulling through without the equivalency hypothesis. According to Nordström, as with me, a redshift of solar spectral lines exists, *but no deflection of light rays in a gravitational field*. The analyses at the next solar eclipse ought to show which of the two versions conforms with the facts. Nothing can be done here at the theoretical level. Next year you astronomers could do theoretical physics a virtually invaluable service in this matter. We will gain reliable information on whether it is correct to continue to generalize relativity theory or whether we must stop at the first step.[14]

As we see, in Einstein's own opinion the only serious rival to his theory at that time was Nordström's scalar theory. In this theoretical scenario Einstein's prediction of the deflection of light in a gravitational field was of particular interest, because his theory predicted this effect in an inalterably specific order of magnitude, while Nordström's theory excluded the effect entirely. Thus detection of light deflection grew to the status of an *experimentum crucis*. As Einstein himself informs us, the effect's mere presence would decide which theory was to prevail: Nordström's more conservative route, which preserved Euclidicity and advocated a scalar gravitational potential, or Einstein's radical solution of adopting non-Euclidean space-time and tensorial fields of gravitation. Though Nordström's scalar theory of gravitation, predicting the propagation of light rays along straight paths, supposedly, as Einstein put it, initially "suggested itself much more," it was based on the *a priori* Euclidean four-dimensional conception of space, which "essentially amounts to superstition." Even Mie's most recent "quite heated polemics" against Einstein's theory[15] only demonstrated the "inadequacies" of the "former point of view based on Euclidean space-time," which Einstein himself had accepted until 1911. Einstein then continues a little sarcastically: "I am glad that my colleagues are at all concerned with the theory, even if for the present it is only with the intention of knocking it down."[16] Further debate on the theory now seemed to be superfluous. At this juncture the crucial experiment would show which path to follow.

When Freundlich recognized in 1913 that the existing photographic plates from earlier solar expeditions could not yield precise enough data on the position of the background stars during solar eclipses (see Chap. 2), he turned his attention to gathering new material. In his correspondence with Campbell and C. D. Perrine of the Cordoba Observatory, Freundlich urged the observers to include a light deflection test among their scheduled measurements during the upcoming expedition to record the solar eclipse on August 21, 1914.[17] When Campbell could not promise this to him definitely in 1913,[18] Freundlich also made last-minute arrangements at the end of 1913 for his own German expedition to the area of total solar eclipse in the Crimea.[19] Einstein was well placed in the Berlin scientific scene to help Freundlich track down funding (see Chap. 5) through his position at the Prussian Academy of Sciences as well as at the Kaiser Wilhelm Institute of Physical Research, the latter of which was later established officially for him in 1917. A few months before Freundlich's departure on this potentially decisive solar eclipse expedition to Russia, Einstein again emphasized his belief in his hypothesis: "I have reconsidered the theory from all angles and can only say that I have complete confidence in the matter."[20] This expedition of 1914 failed, however, because of the outbreak of World War I. Freundlich and his companions, Dr. W. Zurhellen, who was also employed at the observatory, and R. Mechau, a mechanic of the precision optical equipment manufacturer Carl Zeiss Company, were detained in Odessa for many weeks on suspicion of espionage before they were able to take any measurements—their instruments were confiscated.[21] When Freundlich was allowed to return to Berlin on September 3, 1914, his enthusiasm for verifying the general theory of relativity was undiminished. The astonishingly wide publicity devoted to the results of the English solar eclipse expedition at the end of 1919,[22] which roughly verified within the error margin the theory's light deflection prediction of 1″.75, only fueled Freundlich's ambitions. In 1922, at the end of 1925, and again in 1929 he departed on solar eclipse expeditions to test for the existence of light deflection. Yet, although he carefully chose the observation sites on the basis of local weather statistics, cloudy conditions prevented his taking photographs of the field of stars surrounding the eclipsed sun either on Christmas Island on September 20, 1922, or in southern Sumatra in 1926.[23] It was only in 1929 at Takingeun, in northern Sumatra, that Freundlich's hopes of being able to collect his own light deflection data were finally fulfilled. (See Chap. 9.) An overview of Freundlich's research activities related to verifying Einstein's general theory of relativity follows.

End of 1911–October 1912. Examination of available plate data from previous solar eclipse expeditions for evidence of light deflection in the sun's gravitational field. *Result*: plates inappropriately focused.

1912–13. Estimations on the feasibility of daytime observations of stars near the sun. *Result*: too much scattered light.

1913. Binary star analyses: examination of the axiom of the constancy of the velocity of light in the special theory of relativity against Ritz's emission theory of light.

1914. Analysis of Fraunhofer line measurements by Evershed (1913) and Fabry and Buisson (1910) exploring the possibility of gravitational redshift. *Result*: Redshift is present. But already in 1914 Schwarzschild publishes new data that rather speak against the effect.

1914. Expedition to the Crimea exclusively to verify light deflection during a solar eclipse. Aborted by the outbreak of war; the members of the expedition are taken into custody and their instruments confiscated.

February 1915. Refutation of von Seeliger's zodiacal light hypothesis to explain anomalies in the planetary system. In *November 1915* Einstein explains the mercurial perihelion anomaly with his new theory.

1915–. Investigations of fixed star positions for redshift by spectral class. Observed redshifts are correlated with estimated mean mass and radius for the fixed stars. *Result*: Redshift is present. But von Seeliger and Ludendorff accuse Freundlich in the same year of technical errors and "wishful thinking." (See Chap. 4.) Repetition of these investigations: 1919, 1922, 1924, 1928, 1930.

1920–24. Construction of the Einstein Tower, initially specifically for the purpose of verifying the existence of redshift in solar spectral lines. While the construction is in progress, the scope of the analysis is broadened—redshift never actually is confirmed beyond doubt at the Einstein Tower.

1922. Solar eclipse expedition to Christmas Island to detect light deflection.

1923. Solar eclipse expedition to Mexico, also measuring for light deflection.

1926. Solar eclipse expedition to Bengkulen, Sumatra, also for light deflection.

1929. Solar eclipse expedition to Takingeun, Sumatra, also taking light deflection measurements. Finally, weather conditions are good, and observational data are successfully collected. The evaluation yields too large an angle of deflection ($2''.2$ instead of $1''.75$ at the sun's rim; see Chap. 9).

1931–. Reanalysis of data from solar eclipse expeditions of 1922 and 1924 confirms the tendency: light deflection is larger than predicted in the general theory of relativity.

1931–65. Development of a phenomenological theory to explain the discrepancies between observations and the general relativity theory (photon-photon interaction; see Chap. 12).

CHAPTER 4

Statistical Investigations of Gravitational Redshift, 1915–1916

On several occasions from 1911 onwards Freundlich adopted the strategy of reevaluating older data of various types from the fresh perspective of uncovering previously unrecognized relativistic effects. (See the summary in Chap. 3.) He reused solar eclipse photographic plates and, besides these, solar spectral line measurements that had originally been made to determine solar convection currents. Freundlich's idea of recycling older data must be seen in light of the fact that his professional colleagues considered a verification of the relativistic effect of redshift in the solar spectrum to be very improbable because of the interference of other inextricably overlapping effects. The general consensus appeared to be that "a final decision on this question would be possible only on the basis of specialized investigations over many years."[1] In this situation, in order to verify the predictions of the general theory of relativity it was necessary either to perform precision experiments, which at this time Freundlich was not free to do for lack of access to suitable instruments, or to "look for additional important points of attack in this important question."[2] A light source of considerable mass—that is, one with a large gradient of gravitational potential—was necessary to obtain experimentally verifiable gravitational redshift. Therefore, aside from the sun, fixed stars were especially good candidates, because they were, if not comparable in size, often many times larger than the sun. But little was known *a priori* about their motion relative to the earth, so that the problem immediately arose of distinguishing between the redshift resulting from gravitation and that resulting from the Doppler effect. The relative motions of fixed stars away from earth can

TABLE I

Asymmetry of observed line shifts in the spectra of fixed stars,
differentiated according to Doppler equivalence velocities (v) for two
spectral classification types

Observed v (km/sec)	Number of stars		
	Type B–F$_4$	Type F$_5$–M	Total
$> +30$	6	57	63
$+15$ to $+30$	39	104	143
0 to $+15$	170	210	380
0 to -15	86	207	293
-15 to -30	27	89	116
< -30	9	56	65

SOURCE: Freundlich 1915/16b, col. 19.

only be determined spectroscopically: that is, via Doppler redshift. Now a portion of this observed redshift was supposed to be due to an unknown proportion of gravitational redshift. How could the two effects be separated from one another clearly and indisputably?

Freundlich worked out a statistical procedure that he believed could do this. From the data available to him he tabulated the known spectral line shifts of fixed stars, uniformly distributed spatially and corrected for the solar apex. Assuming an isotropic universe, a distribution of stellar velocities should result that is "grouped symmetrically according to positive and negative velocities around the total zero."[3] Yet his results from this initial, rough data examination were otherwise: "The velocities of the stars derived from the measured line shifts rather cluster around a positive value away from zero. This means decidedly more positive velocities result than negative ones; and specifically, it is particularly marked for smaller velocities between -15 and $+15$ km [per second] and for stars of an early spectral nature."[4] Freundlich assembled in "condensed form" the separate observations of Campbell and others, as listed here in Table 1.

If the observed line shifts are interpreted as Doppler shifts, significantly more fixed stars appear to be receding from earth than moving toward it. If one assumed a static cosmos, which was the general assumption at that time,[5] this was very implausible. Freundlich thus assumed that an asymmetrically shifted gravitational redshift, which corresponds to a star system or galaxy expanding at a rate of 4–5 km/sec, was superimposed upon a Doppler shift symmetrically arranged around the value zero.[6] To address the question of whether the result was due to a systematic error[7] Freundlich then discussed the line shifts separately for the various spectral types following Campbell's classification of 1911.[8] (See Table 2.) In order to relate these redshifts to the predictions of the general theory of relativity, Freundlich first had to determine the average mass of fixed stars within these spectral classes. On the basis of the spectroscopic analyses of type B binary stars by Campbell and others, Freundlich set $(m_1 + m_2) = 14m_\odot$, with one solar mass indicated as $m_\odot$. This was based on the additional premise that "in all probability the percentage of spectroscopic binaries among the class B stars [is] large enough to justify extending the data taken for our purposes from binary stars to all known class B stars."[9] If the observed average redshift for class B stars was entirely an effect of gravitation, then obviously the average gravitation potential, or its mass, could be extrapolated. Freundlich arrived at $20m_\odot$, which conformed at least in order of magnitude with the average mass estimate of class B stars derived from the estimated mass of their binary systems. Using the same procedure, he calculated estimates of the average mass of class A stars from binary system shifts as well as gravitational redshift, ending up with four to five times the value of $m_\odot$ for both estimates, which thus happily led to the same orders of magnitude. He saw in the coherence of these estimates an indication of the actual existence of gravitational redshift.

Einstein welcomed this result and rated Freundlich's work in a letter to the Munich theoretical physicist Arnold Sommerfeld (1868–1951) as "surely fundamental."[10] Apart from the loose extrapolation of binary star systems to all stars of one class and the incomplete observational data, the most vulnerable spot in Freundlich's argument was hidden within a formula that determined the relativistic correlation of the observed line shift f

TABLE 2

*Average spectral line shifts for fixed stars in Doppler
equivalence velocities (v), categorized according to
spectral classification*

Type	Number of stars	v (km/sec)
B–Bs	177	4.7
A	177	1.0
F	185	0.1
G	251	0.0
K	751	2.6
M	143	4.3

SOURCE: Campbell 1911; reprinted, e.g., in Freundlich 1915/16b, col. 19.

with density ρ and mass m in units of solar mass $m_\odot$, or solar density $\rho_\odot$. In his first short notice, Freundlich had set $m = f / \sqrt{\rho}$; and after inserting the observed shift f and the estimated density $\rho = 0.1 \rho_\odot$, he determined from this:[11]

$$ m = f \cdot \frac{1}{\sqrt{\rho}} = 15 m_\odot . \tag{1} $$

The astronomer Ritter Hugo von Seeliger (1849–1924)[12] noticed the deficiency in this formula. Relative line shift f and gravitational potential ϕ, radius r, density ρ, and mass m are related according to general relativity theory in the following manner:

$$ f = \frac{\phi}{c^2} \sim \phi = \frac{m}{r} \; ; \quad \rho \sim \frac{m}{r^3} \Rightarrow f = (m \cdot \sqrt{\rho})^{2/3} $$

$$ \Rightarrow m = f^{3/2} \cdot \frac{1}{\sqrt{\rho}} \simeq 65 m_\odot . $$

On June 27, 1915, he passed this corrected formula on to Freundlich's superior, Hermann Struve,[13] the editor of the *Physikalische Zeitschrift*, who in turn informed Freundlich. Von Seeliger's choice of this quite unusual indirect path through Struve to contact a member of his own professional community instead of informing Freundlich personally is explained by the fact that the two had already had differences on another occasion. Freundlich had recently published a paper critical of von Seeliger's zodiacal hypothesis, in which he demonstrated that it could not explain observed anomalies in the planetary system, particularly of Mercury and Venus.[14] In this context the discovery of a hasty error in Freundlich's paper was a welcome opportunity for von Seeliger to strike back. Freundlich's reply to Struve on August 7, 1915, was that he was "very disheartened" that "such a gross and unnecessary mistake could still happen" to him, thanking him for the information and promising to correct it.[15]

This correction appeared in the *Astronomische Nachrichten* in 1916, albeit without naming von Seeliger, with the following comment: "At this point I would not like to forget to add that an error had slipped into the foregoing communication on this question . . . , where a formula that I had initially only used for rough calculations was printed."[16] Freundlich also concedes there that with the change in the formula, the consistency between relativity theory and observation had been damaged, at least if the old estimated values for the density of class B stars were adopted: "If one were to introduce into this formula 0.1 solar densities as the average density of class B stars, as is concluded from an analysis by H. Shapley [*Astrophysical Journal* 38: 158], then the result is not a favorable one. Our knowledge of the density of the stars is not yet remotely sufficient to require that this value apply to all class B stars."[17] With a somewhat vague reference to analyses of average masses in correlation to spectral type by Henry Norris Russell "and others," which conclude that class B stars assume a "middle position," Freundlich then based the new density estimate so critical in the formula to the observed shift as a function of mass at $0.5 \, \rho_\odot$. Thus the "approximately $25-30m_\odot$" reached was then again still acceptably consistent with the value $15m_\odot$ proposed by the theory.[18]

Three issues later, as could almost be expected, Hugo von Seeliger censured not only Freundlich's revision—in which his correction had been incorporated without acknowledgment—but also the whole line of reasoning as well. Freundlich's analysis was publicly torn to bits. On the basis of the corrected formula above for the shift as a function of density ρ and mass m $[m = f^{3/2}(1/\sqrt{\rho})]$, using Freundlich's original figures $f = 4.5/0.6$ and $\rho = 0.1\rho_\odot$, the mass of class B stars comes to $65m_\odot$, and by no means to the value indicated to conform with the relativistic prediction of $14m_\odot$; and as von Seeliger concludes: "the desired conformity has turned into a total failure."[19]

Von Seeliger then continued to show how Freundlich had altered density ρ in order to maintain a value compatible with the mass from the binary star estimate for class B stars. Instead of taking $\rho = 0.1\rho_\odot$ as initially, in his "improved paper" he had applied $\rho = 0.5\rho_\odot$ in order to arrive at $20-30m_\odot$—incidentally "toward the more favorable side of the averaged values."[20] As von Seeliger explained further, in order to arrive at the binary star estimate for class B stars determined by Campbell at $14m_\odot$, a hypothetical density of $2.2\rho_\odot$ would have to be assumed, which appeared to him to be simply unacceptable.[21] These elective adjustments of only indirectly determinable parameters revealed, in von Seeliger's view, "how arbitrary and meaningless the whole calculation is."[22] He concluded:

Thus the result of the whole study is that as long as Ludendorff's and Campbell's figures are retained, it can be established not only that there is no indication of the presence of a gravitation effect, but what is more, that there is only a complete contradiction of the latter. Criticizing the data used is as little the intent of these lines as it is to criticize Einstein's theory. The object is only to present Mr. Freundlich's method of observation and the manner in which he evaluates clarifications that are brought to his attention.[23]

Sloppy though Freundlich's argumentation was, it is nevertheless astonishing how energetically von Seeliger's otherwise rather principled pen slashed away at his opponent. A private letter to Struve that he wrote directly after reading Freundlich's "improved paper" shows the extent of his irritation. According to von Seeliger, the controversial parts of the paper were

scientifically *disingenuous*—to put it mildly—the like of which I have never encountered before in my 40 years of experience in science. Scientific decency, which is otherwise practiced without exception, would have required that the source of the correction be indicated, and professional candor should have prevented such distortions of the real state of affairs. The other salvaging attempts with the obvious tendency to conceal have completely miscarried as well. . . . But in the interest of our science I cannot silently accept Mr. F's outrageous proceedings, and I will bring it out before the public in some form or other, even though I keenly regret charging one of your official employees with such a serious reproach.[24]

The fact that von Seeliger was an outspoken antirelativist says a good deal about this emotional outburst. It was immediately clear to all knowledgeable readers of the notice that despite his explicit statement to the contrary, von Seeliger was attempting to land a stinging blow on the "relativistic fad." This is documented by another of von Seeliger's letters to Struve, in which he gives his reasons for bothering to respond at all to Freundlich: "That was, by the way, only expedient to the extent that I happened to hear that Einstein sets great store in Dr. F.'s line of reasoning. You know that I am extremely skeptical of many of the latest physical hypotheses, and that is why the question under debate seemed to me to be of some interest."[25] Attacks on Freundlich were thus actually aimed at Einstein, and the latter was perfectly aware of this; Einstein wrote to his Munich colleague Sommerfeld in his characteristically humorous tone: "Tell your colleague Seeliger that he has a ghastly temperament. I relished it recently in a response he directed against the astronomer Freundlich."[26] Through his reckless advocacy of relativity theory Freundlich had become the whipping boy for the mainly conservative and antirelativistic community of astronomers in Germany. Cognitive and social elements now complicate this drama between individuals, based on the predominantly critical reception of relativity theory within the scientific community before 1919.

This episode was to have repercussions for Freundlich in his later professional life as well (see Chap. 9), but it also had immediate consequences. Karl Schwarzschild, director of the Astrophysical Observatory at Potsdam since 1909, obtained a negative impression of his former Göttingen student. In a letter to Freundlich's Munich opponent, referring to von Seeliger's second reply, he wrote: "From this it is clear that Freundlich was decidedly dishonest—gravitational redshift absolutely had to be extracted. As you show, nothing of note remains of it. I almost regret it, since I have gradually also taken a fancy to Einstein's theory—Hilbert's version is even finer."[27] Indeed, Freundlich's analysis of older observational data demonstrates how much he was driven by the desire to verify gravitational redshift. Parameters that had not yet been determined definitively, as here the mean density of class B stars, were chosen at will so as to attain an approximate consistency with the predictions of the general theory of relativity. Taken by itself this "fine-tuning" of the variables—a thoroughly common procedure in particle physics to adapt hypothetical models of processes to observation—does not produce credible results in the long run.[28] In order to convince others, Freundlich would have had to justify why he chose to set the relative density at 0.1 and then suddenly a year later at 0.5. This was a tall order, however, in view of the still very inadequate existing observational data. It was generally concluded from Freundlich's results that the effect might possibly be somehow involved, but not that it necessarily was involved. Interestingly enough, several times in his later researches Freundlich repeated this same methodological error of not being critical enough of his own data interpretation.

Nonetheless, Einstein attempted to extract a somewhat more positive finding out of Freundlich's fixed-star data analyses. In his discussion of the foundations of the general theory of relativity of 1916 he added the following cleverly worded comment, simultaneously distancing himself from Freundlich's research efforts: "The existence of such an effect [gravitational redshift] is supported, according to E. Freundlich, by spectral observations of certain types of fixed stars. A final verification of this consequence has yet to be done."[29] In light of the controversies described above over Freundlich's manipulation of the variables, this statement has a distinctly euphemistic ring. When in February of 1916 Struve furnished Einstein with the issue of the *Astronomische Nachrichten* containing von Seeliger's reply to Freundlich and a survey by Hans Ludendorff of the current knowledge on the masses of spectroscopic binary stars, Einstein responded, obviously unimpressed by von Seeliger's scathing conclusions and no less convinced of the validity in principle of the path shown by Freundlich:

Thank you very much for the issue of the *Astronomische Nachrichten*. . . . Seeliger's article tells me nothing new. Ludendorff's I found very interesting, from which I learn how incomplete the available observational data still is. But the good news is that one gets the impression that data leading to a definite decision are bound to come in bit by bit.

The following can be asked in connection with Ludendorff's article: Do weaker stars of the same spectral class also have on average a smaller mass? This could be answered in the case of binary stars, where their mass value can be obtained through averaging. If there should turn out to be a significant correlation, then Ludendorff's approach could prove to be very valuable in the future. As long as the average error exceeds the determined redshift . . . , the result remains completely uncertain. . . .

In any case, I see that Freundlich's result is by no means secure (not even qualitatively).[30] However, credit should be given to Freundlich for having been the first to point out a practicable way to test the question.[31]

Even though Einstein was not fully convinced by von Seeliger's and Ludendorff's objections, through this ugly controversy it did become clear to him how marginalized Freundlich had already become as a result of his overeager attempts to prove the theory. Even Arnold Sommerfeld, one of Einstein's strongest advocates, took von Seeliger's side in advising Einstein to distance himself from Freundlich. Einstein did begin to dissociate himself personally from his intellectual disciple, but he did not break all ties. He explained his position to Sommerfeld as follows:

Freundlich is hardly creatively talented but intelligent and resourceful. His greyhound[32] nature . . . comes in a large part from his pounding heart as he investigates a scientifically important issue. We must not forget that Freundlich had devised the statistical method that permits using fixed stars in addressing the question of line shifts. Although the hasty calculation error has slipped by him and some other things there are greyhoundlike as well (density definition), the overall value of the matter ought not be forgotten because of it. Errors can be corrected and in time always are corrected. The accomplishment lies in discovering a way and smoothing it until it becomes passable.

Seen from my point of view, the affair looks like this. Freundlich was the only colleague in the field [of astronomy] to have supported my efforts effectively up to now in the area of general relativity. He devoted years of thought, and of work as well, to this problem, inasmuch as this was possible beside the laborious and tedious duties at the observatory. What a rascal I would be if now, after the idea has become accepted, I (simply) dropped the man on the consideration that I am now no longer dependent on him. Just put yourself into my shoes. Then you will stop using the watchword "Landgrave be stern."

. . . I simply lack contacts in astronomy.[33]

This and other letters to third parties are not limited to carefully phrased acknowledgments of Freundlich's efforts. There are also some astonishingly deprecating comments about his qualities as a scientist and as a person:[34]

Freundlich is more or less a member of the "greyhound" breed, as defined by a good acquaintance of mine. His way of bolting is also not particularly distinguished. I have known this person's weaknesses for a long time—I have also been more or less irritated by him. . . .

. . . I would *not* choose Freundlich as an intimate friend of mine but would always keep my distance from him either way. And yet I come to the conclusion that if the devil were to unseat all our professorial colleagues whose self-criticism and decency are not above Freundlich's, then the trusty ranks would be considerably thinned. *Horribile dictu*—I would even fear for your informant Seeliger![35]

Following Freundlich's disastrous attempt at finding statistical proof of gravitational redshift in the spectra of fixed stars, Einstein clearly recognized the danger to his own reputation and that of the theory, and dissociated himself from Freundlich in his communications with Struve.[36] How realistic Einstein's fears were is revealed in the following excerpt from a letter of Sommerfeld's to Hermann Weyl, indicating his low estimation of the likelihood of verifying gravitational redshift at that time:

But above all I would like to refer you . . . to the redshift of spectral lines. Up to now no indication of this has appeared. Schwarzschild did not find any; new, careful American measurements at Mt. Wilson also did not. I would be interested to find out from you, when there is a chance, whether redshift is unavoidable in your theory as well. *What Freundlich has published on this as a purported verification of it is more or less a fraud.*[37]

When an influential supporter of relativity theory speaks of "fraud" with reference to Freundlich's experiments, the situation is serious indeed. By 1919 the tone in Einstein's private correspondence had changed significantly with regard to Freundlich.[38] In a letter of December 15, 1919, to Arthur Stanley Eddington, his advocate among English astronomers, Einstein appreciated his having offered a few words of acknowledgment to Freundlich: "He is very diligent, but as a result of material and personal impediments he has not been able to contribute much yet to the theory."[39] In addition, in the tenth edition of his popular pamphlet on special and general relativity (*Über die spezielle und die allgemeine Relativitätstheorie*) Einstein alludes to Freundlich's fixed star data and, without naming him, notes: "In statistical analyses of fixed stars, average line shifts toward the long wavelength end of the spectrum are surely present. But current evaluation of the data does not yet allow a final decision on whether these shifts can really be attributed to gravitation."[40]

A final remark on Freundlich's statistical investigations of fixed stars (which, incidentally, he took up again in 1919, 1922, 1924, 1928, and 1930): verification of the general theory of relativity via this route was very much more complicated than anyone could have anticipated. Both Einstein and Freundlich started from the premise of a static cosmos, in which generally no Doppler redshift was expected to be present in remote light sources. In the first years of applying general relativity theory to cosmological problems, Einstein categorically rejected Alexander Friedman's suggestion of the possibility of obtaining dynamic results from the field equations.[41] However, Edwin Hubble's investigations in the 1920's demonstrated that the cosmos actually is expanding. Faraway fixed stars are moving away from the earth at a speed proportional to their distance.[42] Thus, redshifts in the spectra of fixed stars are obscured by cosmological problems as well as by the effects arising from star development; and until more was known about these factors, it was hopeless to try proving gravitational redshift from fixed-star position data.

At the end of the day Freundlich's controversial research actually achieved the opposite of what he had so enthusiastically intended. Instead of confidence in the existence of an "Einstein effect," there was rather an increased distrust in new simplistic, monocausal explanations. The dominant attitude among theorists and experimenters toward what lay ahead was skepticism regarding conclusions drawn from existing and earlier experiments, combined with a hope that in the distant future quantitatively more reliable experiments would eventually eliminate the disturbing secondary effects. In addition to pressure effects there were also temperature deviations within the spectroscope, the influence of electric and magnetic fields (Stark and Zeeman effects), and the influence of current parameters (pole effect) in the spark and arc spectra used for comparison purposes. All these effects, both those clearly established in the laboratory and other postulated effects (such as the possible influence of anomalous dispersion), had to be eliminated from the "corrected data" before gravitational redshift could be revealed.[43]

CHAPTER 5

Berlin Science Politics:
Support for Freundlich from 1913

Einstein's avid interest in collaborating with an active astronomer in test-ing his general theory of relativity, and his continued consultations with him about alternatives (daytime observations and spectral analysis) even after the failure of the initial verification attempts, matched Freundlich's own dependence on Einstein in executing this research program. Max Planck (1858–1947) and Walther Nernst (1864–1941) had managed to lure Einstein to Berlin in 1914 with an attractive offer:[1] Einstein was free to lecture at the university at will and in October 1917 officially became director of the Kaiser Wilhelm Institute of Physical Research. The sole function of this institute prior to the construction of its offices in the 1930's was the promotion of research.[2] In this capacity Einstein also be-came member of the Prussian Academy of Sciences, taking a salary of 12,000 marks, half of which was paid by the Jewish banker and phi-lanthropist Leopold Koppel (1854–1933).[3]

Einstein used his newfound prestige in Berlin to promote his sole supporter within astronomical circles. The first occasion was at the end of 1913, when Freundlich took steps to find funding for an expedition to Russia to test the light deflection hypothesis during the solar eclipse on August 21, 1914. As soon as he had received the particulars from Freundlich, Einstein took the necessary measures. He applied to the Prus-sian Academy for a grant for the planned expedition through Planck, who advocated the matter "really earnestly" and even undertook to discuss it with the astrophysicist Karl Schwarzschild.[4] But even their support was no *conditio sine qua non*, as he pointed out to Freundlich:

If the academy is not keen on becoming involved, then we will get that little bit of Mammon from private quarters. Right after the academy has informed us of its negative decision, I will with Haber's help touch Mr. Koppel, who, as you know, had parted with the money for my salary as academician. Should all else fail, I will pay for the matter myself out of my little bit of savings, at least the first 2,000 marks. So, after careful consideration, go ahead and order the plates and don't let time slip by on account of money.[5]

In the end Einstein did not have to pay the 2,000 marks out of his own pocket. The Prussian Academy approved the purchase of scientific instruments and photographic plates, thanks to the support of the academy's permanent secretary, Max Planck, as well as that of Walter Nernst.[6] Freundlich also received commensurate contributions from private funders for his expedition.[7] But Einstein's willingness to use his own savings if necessary in order to assure the departure of Freundlich's expedition shows how important the project was to him.[8] Upon learning of the academy's approval of the grant application in March 1914, Einstein immediately wrote to Freundlich: "You can imagine how pleased I am that the external difficulties of your project have now been overcome, so to speak. The fine way in which all the persons involved engaged themselves in this affair is no less gratifying. Planck especially I cannot praise enough. . . . At the *beginning of April* I will be coming to Berlin and am anxious to follow the matter from close at hand."[9] Einstein closed with remarks of a more personal nature, pleased with the prospect of organizing a string quartet with the amateur cellist and future neighbor. The unfortunate outcome of the expedition described above in Chapter 3 did not mar their relationship.

In this episode the first signs of resistance to Einstein's patronage of Freundlich appear during the funding negotiations at the academy. In a letter to Freundlich dated December 7, 1913, for example, Einstein underscored that he would not write to the director of the Royal Observatory in this matter: "*I will not write to Struve.*"[10] Clearly they could expect no support for their plans from Freundlich's superior.[11] Struve's own research was in the spirit of traditional positional astronomy, aimed at compiling as accurate a star catalog as current technology would allow. Hermann Carl Vogel had extended this astrometrical research at Potsdam to include the development of purely phenomenological spectral classifications.[12] Thus Struve could not see the sense in Freundlich's highly theoretically motivated analyses of a pitifully small effect postulated by an incomprehensible theory, and consequently did everything in his power to return Freundlich to his meridian circle measurements.[13] When his assistant became embroiled in a public dispute with Ritter Hugo von Seeliger, a respected Munich astronomer and close friend of Struve's, and, from their point of view, even attempted to hush it up as well (see Chap. 4), Freundlich completely lost face in Struve's eyes. The fact that Freundlich was not sacked as a consequence of the *Astronomische Nachrichten* scandal evoked consternation in Munich. In a letter to Struve of January 26, 1916, von Seeliger exclaimed: "After the past experiences, I am through with Mr. Freundlich. I can only imagine that his 'nervous and agitated' condition has intensified. Otherwise I'd be faced with a psychological enigma. I am very sorry for you that you cannot just get rid of such a man."[14] But Struve also had to act politically, taking into account Planck, for example, who as permanent secretary of the Prussian Academy of Sciences was a powerful friend of Einstein.

This institutional conflict at the Royal Observatory, which had just relocated from Berlin to its new building in Babelsberg in 1913,[15] is reflected in the correspondence. In a letter to Einstein, Schwarzschild pointed to Freundlich's difficulties at the Royal Observatory and intimated that he might possibly be dismissed on the grounds of this conflict, while also raising doubts about his qualifications in handling the instruments. Einstein's response was:

It has never occurred to me to think of an anti-Freundlich clique. It is generally far from my mind to think about such things. Struve's attitude is understandable. He is an old man and no longer has the flexibility needed to delve into new issues. That is why he assumes a negative stance on technical matters, and this negative attitude also extends to Freundlich, who appears to him somewhat as an incarnation of these things. I readily believe that Freundlich on his side has little tact and ability to deal with others, and in general has little psychological appreciation of his fellows, which makes the circumstances even more disagreeable.

I do not take Freundlich for a very great talent, but for a person with a burning interest and a remarkable tenacity. He was the first astronomer to understand the significance of the general theory of relativity and to address the astronomical issues attached to it enthusiastically. *That is why I would regret it deeply if he were prevented from working in this field.* I know from my own experience that the necessary technical skills can be acquired if the requisite understanding and a great interest are combined. Should such deficiencies nevertheless hamper the enterprise, well-intentioned help from an expert could lead to valuable results.[16]

As already mentioned, Schwarzschild's acquaintance with Freundlich dated back to the latter's student years in Göttingen. At that time

Schwarzschild had been professor of astronomy and director of the observatory there between 1901 and 1909. This common background may possibly also explain both astrophysicists' openness toward relativity theory, which was a quite atypical attitude among their German colleagues.[17] However, Schwarzschild, who might well have been able to offer such "well-intentioned" advice, died in 1916,[18] and the altercations between Struve and Freundlich escalated to the point that another solution had to be found.[19]

At the end of 1917 Einstein first attempted to find Freundlich a position as observer at the Astrophysical Observatory, but this plan was stymied by its director. An opinion drafted for an official at the Ministry of Culture, Hugo Andres Krüss, documents Einstein's estimation of the importance and scientific quality of Freundlich's research at that time:

Of all the younger astronomers at New Babelsberg and Potsdam, Mr. Freundlich is, as far as I know, the only one with a solid knowledge of mathematics, celestial mechanics, and gravitation theory. Though considerably less talented than Schwarzschild, he nevertheless had recognized the importance of modern gravitation theories to astronomy many years before the latter did and has worked fervently ["mit glühendem Eifer"] toward verifying the theory along astronomical or astrophysical lines. For seven years the negative stance of his director has made it impossible for him to carry out his research projects directed toward testing the theory. Instead Freundlich was forced year after year to conduct research that any normally gifted 18-year-old boy could be assigned to do after brief training. In the interest of science it would be very desirable and cheaper if he could be employed at one of the two large institutes in such a way that he could, within certain limits, freely choose the subject of his research. The fact that Mr. Freundlich has not yet had the opportunity to acquire practical skills in certain areas of astrophysics plays an infinitely small role in comparison with the favorable circumstances mentioned—this deficiency can be overcome in a short time. It is quite inconsequential to the plan whether Mr. Freundlich ought to be employed at New Babelsberg or at Potsdam; technically speaking, Potsdam would be suitable in the first place, because testing gravitational theory is regarded as an astrophysical task.[20]

When these various possibilities fell through,[21] Einstein finally exercised his power as director of the Kaiser Wilhelm Institute of Physical Research: upon agreement with the advisory board, which tended to meet at Einstein's apartment before the completion of the institute building, Freundlich signed a contract appointing him the first and sole staff scientist at this institute starting January 1, 1918, "in the interest of conducting experimental and theoretical astronomical investigations to test the general theory of relativity and related issues."[22] This contract placed Freundlich in the fortunate position after the war of finally being able to dedicate himself entirely to his own research.[23] At this point one might wonder how fortunate this really was. Was it not rather research for his patron, Einstein, who was immediately interested in the verification of his theory? Had it not been with Einstein's encouragement that, upon receiving Pollak's unassuming inquiry, Freundlich had started to neglect his other duties in order to pursue the intricate problem of detecting effects at the limit of the measurement capabilities of the time? Had he not succeeded in organizing the expedition to Russia only through pressure from Einstein, ending up a prisoner of war there? His advocacy of the unpopular theory of relativity had turned Freundlich into an outsider at his own place of employment. He was constantly admonished to return remorsefully to his cataloguing work, or preferably to find an occupation that "better suited his abilities."

Now Einstein had provided him with a position that gave him a bit more freedom of action; but he was dependent nonetheless, dependent upon the continued good will of his patron, upon whose support the success of his projects rested. This is documented, for example, in the correspondence between Einstein and Planck on the details of Freundlich's contract, the first of its kind for the institute. Among the terms they discussed was the requirement that Freundlich should submit regular reports on each comprehensive analysis to the Kaiser Wilhelm Society's board of directors "at the latest upon completion of the preliminary research" but at least once annually; he would further be required "to take into account wherever possible any change suggested by the board of directors along with any other suggestions." The correspondence also considered whether Freundlich should have to obtain the approval of the board prior to publishing his results. In Einstein's opinion this requirement went too far, and he suggested striking this second paragraph of the draft contract. Planck responded:

Naturally I fully support your intention to omit §2 of the contract with Freundlich. . . . But I would still like to emphasize as a matter of principle that in general in other contracts of this type I would set some store in a provision like the one in §2. This is because I could very well imagine a case in which someone who has been supported by the Kaiser Wilhelm Institute does not come up with anything decent in the end and then publishes his worthless results on his own initiative with reference to his connection to the K. W. Institute and in the hope of promoting himself with it. This could put us in an awkward position, and that is why we should try to protect ourselves in general from such incidents.

But in Freundlich's case nothing is in jeopardy to the extent that you have him in hand, so to speak.[24]

Freundlich certainly was kept under tight rein. Because of a provision in his contract that required him to seek approval for any expenditure over 50 marks per month, he soon gained the reputation of constantly plying for money. One illustration of this is Planck's letter to Einstein of July 20, 1919, in which he writes: "The whiner [*Schmerzenskind*] Freundlich needed money, and fast: First 300 marks for his microphotometer, and then a cost-of-living bonus to his salary."[25] There were various attempts to change this unedifying situation, such as applying for a position in Vienna; these failed, and Einstein could do no better than console Freundlich with the words: "Everything will straighten itself out somehow. You have not got nearly enough padding. Your nerves lie completely exposed, without any protective layer of fat!"[26] In a report that Freundlich sent in 1918 to his former mathematics professor Felix Klein at Göttingen, additional disadvantages to his new position also become apparent. It was a three-year temporary position, renewable for two years, leaving his future prospects completely up in the air since he had irreversibly severed all ties with his previous employer:

In the near future I will probably have hardly any opportunity to work on the theoretical aspects of these [relativity] questions, since I am devoting myself entirely to the experimental testing of the theory now that the Kaiser Wilhelm Institute of Physical Research has made me independent for some years, enabling me to leave my position at the Royal Observatory. At the moment I am working exclusively at the astrophysical institute in Potsdam and am developing a number of methods, first to determine the gravitational shift of spectral lines, should they exist, and then also to verify the deflection of light in a gravitational field.[27]

At the end of 1919, Freundlich launched an appeal directed at potential financial donors that was to succeed at last. It called for the creation of a special fund for the astrophysical verification of the general theory of relativity. (See Chap. 6.) In this way he hoped to procure for himself some kind of long-term position as a senior scientist. Freundlich's later tendency to refer to the Einstein Tower frequently as his own institute, disregarding the contributions of all others involved in its establishment, was based on this petition.[28] But his total dependence on Einstein's sup-

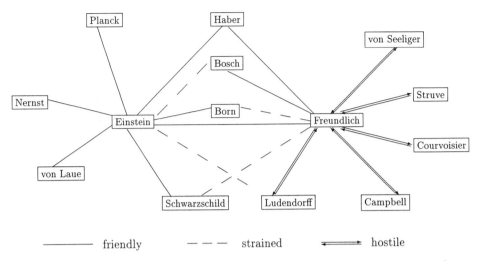

Fig. 6. Interaction diagram comparing Einstein's harmonious connections and Freundlich's confrontational stance.

port cannot be forgotten. Freundlich's controversial position already at this stage of his career is illustrated in Figure 6.

Freundlich's relations with Einstein were also beginning to deteriorate by this time. (See Chap. 8; see also Chap. 10 on Ludendorff.) The growing number of enemies Freundlich was making begs the question of whether or to what extent his combative character led to a polarization in regard to relativity theory, and whether many of his well-intentioned efforts in promoting it rather produced the opposite effect.

CHAPTER 6

Astrophysics at Potsdam and Elsewhere

Around 1860, the spectroscopic researches of Kirchhoff and Bunsen opened the way for astronomers to exploit new methods not only of registering light from distant stellar bodies, but also of analyzing it. Through Doppler shift, comparisons of spectra from stellar sources against spectra generated on earth revealed information about the chemical composition of the stars and their speed relative to earth. More refined procedures, for example those that implemented the Zeeman and Stark effects discovered in 1896 and 1913 respectively, also added details about the physical conditions present in the luminescent atmospheres of stars. Thus the new discipline of astrophysics came into being, which subjected the electromagnetic radiation emitted from astronomical objects to physical methods of analysis.[1] Two sciences that had previously been completely separate—positional astronomy, which catalogued and classified, and physics, which analyzed and explained—had joined forces.[2] In 1871, with the support of Wilhelm Foerster (1832–1921), then the director of the Berlin Royal Observatory, it was decided to construct "an observatory on a favorably located spot near Berlin, equipped with instruments for the direct spectroscopic and photographic observation of the solar surface." The Telegraphenberg on the outskirts of Potsdam was chosen, and Hermann Carl Vogel (1841–1907) became the first director of the Potsdam Astrophysical Observatory (Astrophysikalisches Observatorium) established in the autumn of 1876.[3]

This new disciplinary alliance found outside support as well: in the Prussian Academy of Sciences Emil Warburg, a member of the commit-

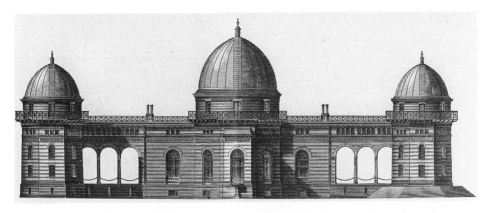

Fig. 7. The Astrophysical Observatory, Potsdam. (From Scheiner 1890: pl. 5)

tee responsible for finding a successor to the late director of the Astrophysical Observatory in Potsdam, Karl Schwarzschild, who had died in 1916, suggested appointing a physicist.[4] He was supported by Einstein, another member of the committee, but the latter had not yet become powerful enough in Berlin to counter the astronomers' categorical opposition to the idea. Gustav Müller (1851–1925)[5] was nominated, a pupil of Wilhelm Foerster's who had been quite influential in the foundation of the Astrophysical Observatory as well as the Physikalisch-Technische Reichsanstalt in 1887 and whose "familiarity with the local situation, aside from his scientific achievements," spoke for his selection. Müller had directed a photometric survey of the northern skies between 1886 and 1906, cataloguing a total of over 14,000 stars of magnitudes up to 7.5. This is a prime example of the research conducted at Potsdam, which demanded "extraordinary diligence, conscientiousness, and dutifulness" but not necessarily much originality. In an obituary Müller was characterized as follows: "He was not a man of surprising new ideas. . . . Müller was thoroughly an astronomer of the old, exact school. Therefore the more recent developments in astronomical research did not always necessarily appeal to him, but he was far from flatly rejecting them either."[6]

Despite his entirely different character profile, Müller seems to have gotten along quite well not only with Einstein, toward whom he was positively disposed, but also with Freundlich, presumably because of his universally acknowledged tolerance.[7] When Müller retired in 1921, the physicists on the appointment committee, Emil Warburg, Max Planck, and Walther Nernst, tried again even more resolutely to carry through one of their own.[8] Einstein, who had no interest in the position himself, and his good friend Max von Laue (1879–1960),[9] a theoretical physicist, were suggested. The astronomers again protested, but a compromise was finally reached. The Prussian Academy was to nominate both von Laue and the astronomer Hans Ludendorff[10] jointly as first choice, on the following justification:

The development of modern astrophysics has brought along with it ever closer ties between astronomy and physics. Astronomers cannot pass over in silence the theoretical and experimental research by men like Einstein, Eddington, Michelson, and the rest, which has led to extremely significant accomplishments; and on the other hand, physicists have a great interest in advances in astronomy. Only through the intimate collaboration of the best minds in both fields can we expect any finished results.[11]

The Ministry of Culture did not relish the prospect of appointing two directors at the same level, however, with Ludendorff performing all the managerial duties. Finally, Ludendorff received the appointment but was obliged to report regularly to a newly established board, to which his predecessor as well as the theoretical physicists von Laue, Einstein, and Planck were elected. In addition to performing general supervisory functions, the board's purpose was to advise the astronomers on physical questions. In order to guarantee the "best possible efficiency" of this advisory committee and to "facilitate very frequent meetings," its membership was to consist "only of persons who reside in the Berlin suburban commuting area."[12] Max von Laue became the presiding chairman of this advisory board, and a representative of the Physikalisch-Technische Reichsanstalt, which focused on metrology and instrument calibration, was also added: until 1925 the antirelativist optics specialist Ernst Gehrcke (1878–1960), who was succeeded by the spectroscopist Friedrich Paschen (1865–1947). Between 1922 and 1933 the members of the Astrophysical Observatory's advisory board were:[13]

February 27, 1922–March 31, 1925:	Albert Einstein, Max von Laue, Gustav Müller, Walther Nernst, Ernst Gehrcke
April 1, 1925–March 31, 1928:	Einstein, von Laue, Hans Ludendorff, Müller (d. 1925), Nernst, Friedrich Paschen
April 1, 1928–March 31, 1931:	Einstein, von Laue, Ludendorff, Nernst, Paschen

Additionally from June 1929:   Erwin Schrödinger
April 1, 1931–November 1933:   Einstein, von Laue, Ludendorff,
                                       Nernst, Paschen, Schrödinger
November 1933:   Dissolution of advisory board.

As astrophysics developed, difficulties arose not only in institutional and staffing questions, but also with regard to instrumentation. Instruments were first to reflect the fusion of the different observation methods of astronomy and physics. For example, a simple physical tool, the prism, was attached to the end of the tried-and-true telescope to analyze intercepted light.[14] But telescopes became increasingly cumbersome as they accumulated more and more astrophysical apparatus, and the precision required in mounting and adjustment was consequently impaired.

The first to address this increasingly pressing problem with a radically new observatory design was the American George Ellery Hale (1868–1938).[15] Several large observatories were built on his initiative in the United States, each of them a milestone in the history of astrophysical observatories. In 1897 the Yerkes Observatory was inaugurated in Williams Bay, Wisconsin, 120 km northwest of Chicago, equipped with a huge refractor carrying lenses 102 cm in diameter—the largest refractor ever built.[16] From 1904 onwards, the Carnegie Institution and other private donors funded the creation of the Mt. Wilson Observatory near Pasadena, California, which was initially mainly intended as a solar observatory.[17] The first telescope in operation there was the Snow telescope. (See the upper photographs in Fig. 9.) It directed the image of the sun through a double-mirror system (called a coelostat; see Figs. 21, 33) along a horizontal shaft for further analysis.[18] Temperature changes and drafts at ground level often had a noticeable and disturbing impact on the Snow telescope's performance, however; and Hale decided to try out another arrangement of mirrors, lenses, and analysis apparatus. Near the Snow telescope another telescope was erected exclusively for astrophysical observations: a "tower telescope" was built between 1904 and 1908. It was equipped with a coelostat at the tip of a 60-foot tower that directed sunlight vertically downward into a ground-level laboratory. The lenses, made of particularly thick glass, were cooled by electrical fans during measurement to limit distortion caused by any heating up of the optics. (See the lower photograph in Fig. 9.)

When the merits of this new type of telescope became apparent—for example, through the discovery of magnetic fields in solar spots in 1908—

Fig. 8. Telescope with a spectroscope attached, at the Astrophysical Observatory, Potsdam. (From Vogel 1900: pl. 9)

a 150-foot tower telescope of the same basic design soon followed in 1912, with subterranean laboratories for improved maintenance of the constant temperatures required.[19] Thus both telescopes erected at the Mt. Wilson Solar Observatory are positioned vertically, with the lens installed at the tower summit. Only the coelostat in the dome is adjustable; it is connected to a tracking system controlled by a pendulum clock on sidereal time to move against the earth's rotation at half the angular velocity. Through this design, the light received is reflected into the instrument constantly from the same orientation. Another trademark of the observatories founded by Hale is the excellently equipped laboratory associated with each of them. These were provided with electromagnets to measure Zeeman spectral line splitting, electrical ovens to produce emission spectra of luminescent gases at high temperatures, and high-resolution spectral analysis instruments.

At the Mt. Wilson Observatory Hale conducted research on magnetic fields in solar spots, which involved analyzing the minute details of Zeeman spectral line splitting. Arthur Scott King (1876–1957) studied the temperature dependence of spectra, using modern electrical ovens. Walter Sydney Adams (1876–1956) worked together with others on center-limb variations, peculiar shifts in the spectral lines of light originating from the sun. Another person to work here was Charles Edward St. John (1857–1935), the internationally acknowledged expert on high-precision measurement of all types of shifts occurring in solar Fraunhofer spectra.[20]

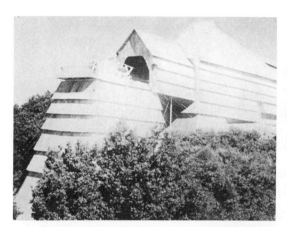

Fig. 9. Early astrophysical observatories at Mt. Wilson: *above*, the Snow telescope (1905) and its coelostat mirrors; *below*, the first 60-foot tower telescope (1908). (From Hale 1915)

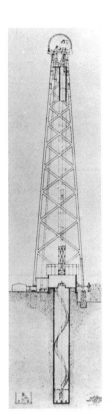

Fig. 10. The 150-foot tower telescope at the Mt. Wilson Observatory: *above*, photograph and diagram; *below*, laboratory interior, showing spectroheliograph. (From Hale 1915)

The excellent equipment at this astrophysical center prompted German scientists to conclude that they also must obtain these new instruments as soon as possible in order not to lose ground completely to the Americans: "You can only have complete admiration and be a bit jealous when you hear about the magnificent equipment that the Mt. Wilson Observatory provides for such researches."[21] But intention alone was not enough. Freundlich worked actively to drum up the necessary funding for "his own" institute; but because he had become persona non grata in so many quarters (see Fig. 6), he moved under Einstein's cover, using Einstein's connections:

I cannot be so demanding of the gentlemen of the Kaiser Wilhelm Society. At the moment the director of the [Buna?] works,[22] a very influential person, has taken the matter in hand and will ask Mr. von Siemens to donate the electrical system (60,000 marks). The director of the Potsdam institute submitted my memorandum to the ministry yesterday together with a copy of the sketch, and he informed me today that he found that my plans had received a surprisingly positive response. We hope to get a grant of 30,000—50,000 marks from the ministry.[23]

But the memorandum mentioned in this letter, which had been submitted by Gustav Müller, became caught up in bureaucratic red tape.[24] When at the end of November 1919 daily newspapers reported that the two English eclipse expeditions led by A. S. Eddington and A. Crommelin had confirmed Einstein's prediction of the existence of light deflection, many German scientists indignantly questioned their country's lack of involvement. Freundlich decided to capitalize on the propitious moment.[25] He drafted his "Appeal for the Einstein Donation Fund [Einstein-Spende]," which was sent in December 1919 to various generous donors in industry and business, requesting a contribution to enable the Astrophysical Observatory to become involved in the experimental verification of the general theory of relativity. Rampant inflation has left its mark on this document. The typed draft of the appeal sought 300,000 marks; the figure was subsequently corrected to 500,000. In the final printed version, however, the numeral 1 has been inserted in front of the sum, raising it to 1.5 million marks![26] The appeal also deftly drew attention to the foreign competition:

The academies in England, America and France have recently set up a commission, which excludes Germany, to establish actively experimental bases for the general theory of relativity. It is an obligation of honor to those who are concerned about Germany's cultural standing to come up with whatever funds they can afford in order to enable at least *one* German observatory to work directly with its creator in testing the theory.[27]

TABLE 3

Excerpt from the list of contributors to the Einstein Donation Fund, 1920

Contributor	Amount (marks)
Baden Aniline and Soda Factory (BASF), Ludwigshafen	24,000.
Dye (Farben) Works, previously Meister, Lucius & Brüning, Höchst	24,000.
Dye Factory Main Cashiers, previously F. Bayer & Co., Leverkusen	24,000.
Carl Zeiss Foundation Jena[a]	20,000.
Robert Bosch, Stuttgart	20,000.
F. Herrmann & Co., Luckenwalde	20,000.
Anonymous	20,000.
Mendelsohn & Co., Berlin	20,000.
Otto Wolff, Berlin	20,000.
Berlin Trade Company (Handelsgesellschaft)	15,084.40
Wolf Netter & Jacobi, Berlin	10,000.
Grand Lodge for Germany, Berlin	10,000.
Leopold Cassella Firm & Co., Ltd., Frankfurt	10,000.
Foreign Trade Office for Iron and Steel Products	10,000.
Anonymous	10,000.
Stock Co. for Aniline Manufacturing, Berlin	9,000.
Chemical Factory Griesheim-Electron, Frankfurt	6,000.
United Lace and Bobbin Industry	6,000.
Siemens & Halske Stock Co., Schramberg[b]	5,000.
Association of German Machine-Building Establishments	5,000.
General Electricity Company (AEG) Secretariat, Berlin[b]	5,000.
Junghans Brothers Stock Co., Schramberg	5,000.
Bank director Dr. Dannenbaum, Berlin	5,000.
Max M. Warburg, Hamburg	5,000.
Adolf Pitsch, Berlin	5,000.

SOURCE: Main Archive of the Academy of Sciences, Berlin, Astrophysical Observatory holdings, file no. 147.

[a] [Original note:] "The C. Zeiss Company has donated optics worth 300,000 marks and is building the planned facility, a tower spectrograph, at cost. The firm Schott & Partners is delivering all the glass under the same conditions."

[b] [Original note:] "An additional contribution to the fund is promised by the large electrical companies in the form of the delivery of the electrical system under special terms, which corresponds to a multiple of the indicated amount."

Within a year this Einstein Donation Fund collected the handsome sum of 350,000 marks from both private contributions and German industrial associations mobilized by the Reichsverband der Deutschen Industrie ("Imperial Federation of German Industry").[28] This association's managing director, Dr. Rudolf Schneider, was appointed to the newly established board of trustees that was to administer the monies donated. Material support from the precision optics company Carl Zeiss, for example, and the Schott glassworks was also significant. The optics and instruments were offered at production cost, and Freundlich was entrusted with a particularly valuable lens initially on loan for a number of years.

The list of contributors from the Einstein Tower files gives an idea of the types of donors involved. (See Table 3.) Government officials decided in the end not to exempt contributions to the Einstein Donation Fund from the normal gift tax.[29] When in May 1920 the Prussian government had still not approved a state grant of 150,000 marks, Freundlich pressed the politicians with another nationalistic argument: something had to be done "to prevent German science's being excluded from the further development of this important field of research—especially considering that not only is the conception of the new ideas the intellectual property of a German, but also, it was in Germany that the first attempts to test it through experiment were made and its consequences pursued."[30]

The costs and expenditures involved in the construction of the Einstein tower telescope up to October 1921 are summarized in Table 4.[31] Why the Einstein Donation Fund deposit shrank so drastically, by over 200,000 marks, between October 1921 and 1922 is not clear. As 1921 taxes had already been incorporated into the first amount, this reduction may have involved retracted donations. For comparison, by the end of 1921 the private organization for the promotion of German science founded in 1920, the Notgemeinschaft der Deütschen Wissenschaft, received 75 million marks (converted into gold marks: 1.6 million) in

TABLE 4
Balance sheet of the Einstein Tower, 1922

Item	Amount (marks)	
	Dec. 1919–Oct. 1921	Grand total, 1922
Income		
Einstein Donation Fund	1,410,582.96	1,190,000.00
Prussian government grant	150,000.00	200,000.00
SUBTOTAL	1,560,582.96	1,390,000.00
Expenses		
Tower telescope construction[a]	794,675.68	850,000.00
Instruments and optics[b]	73,421.13	220,000.00
Electrical laboratory facility	14,387.80	260,000.00
Shipping, wages, printing	44,619.30	45,000.00
SUBTOTAL	927,103.91	1,375,000.00
BALANCE	633,479.05	15,000.00

SOURCE: Main Archive of the Academy of Sciences, Berlin, Astrophysical Observatory holdings, file no. 147: balance sheet of the Einstein Tower, Oct. 1921, and report to the board of trustees of the Einstein Donation Fund by Erwin F. Freundlich, 1922 (doc. 147).

[a] The architect's fee (65,000 marks), paid to Mendelsohn in October 1921, is included in the construction costs.

[b] Most of this expense involved an advance payment to the Zeiss Company of 66,667 marks for the installation of the tower spectrograph, to which 23,250 rentenmarks was added after the currency devaluation of 1924.

contributions; by the end of 1922, a total of approximately 100 million marks.[32]

Because of the rapid inflation, quick financial decisions now became vital to prevent the funds collected from being reduced to a fraction of their worth. In September 1920 the ministries of culture and of finance finally reached agreement on the amount to which the state was willing to commit itself: 200,000 marks.[33] With the question of finances settled, Freundlich was now able to organize the construction of the research facility.

CHAPTER 7

Erich Mendelsohn and the Tower Telescope Design

On April 24, 1920, Einstein signed a general power of attorney on behalf of the Einstein Donation Fund granting Freundlich the practical authority necessary for the tower project: "I grant herewith general power of attorney to Dr. E. F. Freundlich in New Babelsberg in all matters concerning the construction of the tower spectrograph to be built on the premises of the Astrophysical Institute in Potsdam." Written in the unmistakable hand of Erich Mendelsohn, this document bears witness to the close collaboration between the theoretical physicist, the astronomer, and the architect in the building stage of the new Einstein Institute.

Erich Mendelsohn (1887–1953)[1] was born in Allenstein, in eastern Prussia; the second-youngest of six children in a Jewish family of modest means. After his elementary schooling, he attended the preparatory school (*Gymnasium*) in Allenstein, and thereafter in 1907, he began to study economics at Munich University. In 1908 he changed to architecture, studying at the Charlottenburg Polytechnic in Berlin between 1908 and 1910, and at the Munich Polytechnic from 1910 to 1912, where he took his degree under Theodor Fischer. In the following two years he received no commissions and earned his living in Munich by designing posters, stage sets, and shop window displays. Toward the end of his Munich stay he came in contact with artists of the expressionist school known as the *Blaue Reiter* (Blue Rider) and participated in an expressionist theater project. When World War I broke out he moved to Berlin, initially exempt from military service because of his poor eyesight.

The cellist Luise Maas (1894–1980), who became Mendelsohn's wife

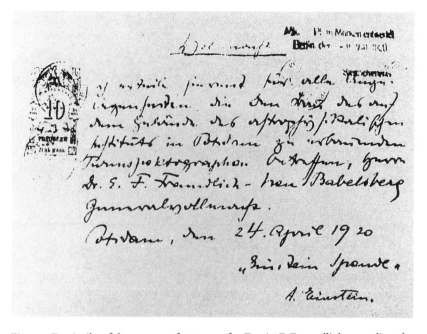

Fig. 11. Facsimile of the power of attorney for Erwin F. Freundlich regarding the Potsdam tower spectrograph, written in Erich Mendelsohn's hand and signed by Albert Einstein for the Einstein Donation Fund. (From State Museums of Prussian Cultural Holdings, Art Library, Berlin, Mendelsohn Papers [hereafter "KB"], Einstein folder, by permission of Prof. B. Evers)

in 1915, was the one to introduce the young man to Erwin Freundlich around 1914. She had gotten to know the astronomer through Käte Hirschberg (1888–?),[2] a friend she had met in Königsberg in 1909. Käte, a former student of the Belgian architect Henry van de Velde, had met Erwin in 1910 and married him three years later. The two couples had many interests in common, ranging from music to art, as their correspondence illustrates. Letters from Freundlich after 1917 are preserved among Mendelsohn's papers together with drafts of some of Mendelsohn's responses. The letters reflect the political situation and their authors' wartime experiences; they also provided Mendelsohn a source of cultural entertainment while he was serving at the front as a sergeant in the army. Freundlich discussed literature and sent him books to read, for example, and reported on the activities of a Babelsberg string quartet, as Freundlich himself was an amateur cellist. For their part, Mendelsohn and his wife procured for their undernourished city-dwelling friends during the war years culinary tidbits like semolina, noodles, rice, eggs, and once even "real cake!"[3] Through his friendship with the Mendelsohns, Freundlich also came into contact with expressionists. In a letter to Mendelsohn in August 1918, he describes a visit to a member of the artists' group *Die Brücke*, the painter and graphic artist Max Pechstein (1881–1955): "I spent last Sunday afternoon and evening with the Pechsteins in their new apartment. They have rented a studio in a large building built by Messel on Kurfürsten Street." After describing the dimensions and layout of the apartment in detail, he continues:

I was again very impressed by Pechstein. He is amazingly amiable and sensitive. His way of responding to creative stimuli is so very naive, such as you would really expect only from a primitive person. I was struck by his painting as well—to be honest, it was the first time in my life that I took away with me any kind of lasting impression from paintings. It suddenly became clear to me how the colorful artistic experience can become quite primordial in an object as well, how necessary abstractness in an object is, and on the other hand, how nevertheless the experience is not possible without objects. This possibility of gaining a personal relationship with a picture came to me more powerfully and more suddenly than I would have ever expected.[4]

Freundlich thus acquired a sense for architecture, painting, and spatial design through these contacts. Pechstein's own impression of the astronomer is captured in a series of woodcuts that he completed in 1918 (see Fig. 12), including a portrait of him playing the cello.

Freundlich's correspondence in May 1917 also mentions his budding plans to extricate himself from his thorny position among the Berlin astronomers. In a letter dated May 18, 1917, Freundlich speaks of "new opportunities" and the possibility that he might take over the direction of an observatory in Vienna. The observatory theme inspired Mendelsohn, at that time serving at the Russian front, to jot down some sketches in a small pad. The earliest such sketches to have survived illustrate the steady development of his architectural form, away from his theater designs in particular. The final design seems as if it had flowed naturally from pen to paper with a momentum of its own. In a letter to his wife dated June 24, 1917, Mendelsohn talks of a union of "tellurian and planetary" elements in these visions.[5]

Freundlich received a few of these loose sketches in the mail, but he had doubts about the applicability of the inventive designs to real structures.[6] Mendelsohn did not deny his ignorance regarding the practical aspects of a working observatory and was open to constructive criticism:

Fig. 12. Woodcut portrait of E. F. Freundlich by Max Pechstein, 1918. (From KB, by permission of Prof. B. Evers)

"I totally agree with your objections to the technical difficulties of the observatory sketches. As 'visions' they have become observatories quite coincidentally, and as such do not feel bound to rigid functionalism."[7] Two months later, though, he wrote:

Until I have the privilege of proving the functionality of my work through its realization I can only state the basic idea somewhat programmatically, whereby the distinct vibrations are usually smoothed over completely.

My sketches are data, the fixing of the contours of a sudden vision.

Their architectural nature makes them appear immediately as a whole and is intended to do so. . . .

Now, every artistic impression is based upon its own energetic escalation.

Segments line up, bow, jut out.

Only the interplay of their energies can transmit the overall tone.

The enhancement of mass and surface forms a part of this.

Mass enhancement means:

Gradation–reclined–inclined masses

Motion–tense–relaxed (iron) masses

[Motion–] supporting–carrying (stone) [masses]–structure!

Balanced mass is rhythmic mass.

Surface enhancement means:

Function–closing–opening surfaces

Richness–uniform–dispersed surfaces, ornamentation!

Balanced surface is rhythmic surface.

Rhythmic relativity is character, harmony.

Both gain life in the outline, both enable us to decipher the outline.

This is an avowal of the fundamental law of all architecture.[8]

In his cryptic, almost poetic style Mendelsohn not only elaborates on his sketches but also reveals his conception of the Einstein Tower. (A selec-

Fig. 13. Mendelsohn sketching, ca. 1920. (From KB, by permission of Prof. B. Evers)

Figs. 14 and 15. Fig. 14 (*left*), early observatory sketches by Mendelsohn, 1917–19. Fig. 15 (*right*), three sketches by Mendelsohn of the Einstein Tower from about 1920. (From KB, by permission of Prof. B. Evers)

tion of the sketches were exhibited in 1919 in Paul Cassirer's gallery in Berlin and then elsewhere under the title "Architecture in Iron and Cement.")[9]

Freundlich's letter dated July 2, 1918, written after his losing the assistantship at the Royal Observatory, traces the feverish beginnings of the new institute:

I am contemplating a project of building a little institute for my own research, once the influential individuals at the Kaiser Wilhelm Institute have approved my plans and I have been explicitly requested to submit a memorandum with blueprints. . . . The director of the Potsdam Observatory has offered me a very suitable plot of land on his observatory's premises as a building site. If it can be arranged, I will try to have you prepare the blueprints for the exterior design, although it will not be a very rewarding job for you. I have considered the matter as follows:

A cement tower 15 meters high is capped by a small dome 1.5–2.0 meters in diameter. The tower has a double wall; that is, its outer shell surrounds a completely isolated chimney with an unobstructed opening of perhaps 500 mm and a wall thickness of 500 mm. A coelostat (heliostat) stands on this chimney, which with the aid of its mirrors reflects the image of the sun vertically downward into a subterranean laboratory, from there onto a horizontal cement foundation (isolated) on which are mounted a slit, camera, and diffraction grating to produce the spectrum. The underground laboratory is about 15 meters long and only as tall and wide as is absolutely necessary, since it must be maintained at a constant temperature. A number of rooms border on the laboratory, opening into it via small windows or double doors, namely (1) the room for the electrical furnace, (2) room for electric arc lamps; both individually well ventilated and provided with electrical outlets or air pump connections. Additionally there is a darkroom and a workroom with an exit to the world above. Only a small house is at the foot of the tower, with perhaps one or two rooms and leading down to the laboratory. My sketch should naturally be regarded only schematically.[10]

The small sketch in this letter reveals how much Freundlich's physical layout for the "little institute" imitates Mt. Wilson's. In 1918 he had thoroughly considered among others the construction Hale had opted for in realizing his 150-foot tower telescope. (See Fig. 10.) There the laboratories had been placed underground, well below the tower, to optimize temperature isolation.[11] Freundlich's letter to Mendelsohn also describes the alternatives in more detail:

At the moment I am still undecided whether I should not give preference to another plan that I had dropped at first. In it the concrete tower remains just as in the above sketch, except that there is a light shaft underground of perhaps 6–8 meters as an extension of the chimney. On the floor of the shaft stand the diffraction grating and camera lens, which reflect the spectrum almost vertically upwards, so that the slit, which cuts the solar image to a narrow band, is located directly below the camera, which is on a table; and all the observations are made in a small house at the foot of the tower. Naturally the tower must be capped by a dome and partitioned so as to provide enough room. This project may well be cheaper and in some respects more favorable as well, since the vertical arrangement of the parts involves the least stress. The advantage is that the grating at the bottom of the shaft cannot be reached directly and the above-ground observation room is not as well insulated thermally. As I said, I am not yet quite sure which of

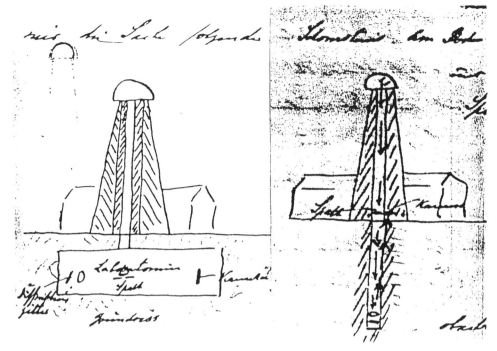

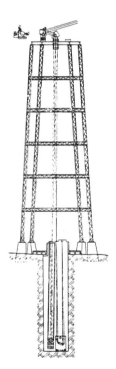

Fig. 16. Two sketches from Freundlich's letter to Mendelsohn dated July 2, 1918 (*above*), and (*left*) diagram and photo of the 150-foot tower telescope at Mr. Wilson. (From KB, by permission of Prof. B. Evers)

these two possibilities is more advisable. Would you have the time and the inclination to make a drawing? Something nice can obviously be made of such a little project as well. Well, a month or two is bound to go by before this thing is built, and I would be very pleased if you could draw up the plans.[12]

Freundlich finally decided on the better accessibility of the horizontal arrangement, which was then realized in Potsdam (see Figs. 21, 32–36): in this arrangement two coelostat mirrors measuring 85 cm in diameter and 17 cm thick are installed in a rotatable dome with an interior diameter of 4.5 meters. The mirrors direct the light vertically downward to the lens telescope, which measures 60 cm in diameter and has a focal length of 14 m (approximately the height of the tower).[13] From there another mirror allows a 13-cm image of the sun, for example, to fall horizontally onto the wall separating the workroom from the darkroom. (See Fig. 36.) The observer can adjust the precision slit in the panel of this wall to select a section of the solar image. Both edges of the slit are highly polished, and the slit itself is slightly tilted, so that during long exposures a portion of the incident light is reflected back into the lens of the smaller aiming telescope, which is attached to an isolated pedestal in the laboratory together with the redirecting mirror. All the necessary controls are ready at hand for a seated observer to position a desired segment of the intercepted light to pass through the slit. The main portion of the unimpeded light then arrives in a thermally insulated darkroom equipped with a prism spectroscope (with prisms 320 mm long and 145 mm high, with a refractive angle of 30° and a dispersion of 2 Å/mm for light of average wavelength [4,000 Å], set by the Schott Company in Jena and cut by the Zeiss optical works), and a grating spectrograph (of 100,000 grooves and a grating constant of 600 grooves per mm), which operates in autocollimation mode. This means that the light rays pass through a lens with a focal length of 12 m, which "combs" the light into a parallel beam. This beam is directed onto the grating, and the spectrally decomposed light is reflected back from the grating through the same lens onto the receivers. The two beams travel on slightly offset paths so as not to interfere with each other. Both instruments can be operated using the controls in the workroom at the slit panel (see Fig. 36), which eliminates the necessity of entering the darkroom during measurement. The light is thus reduced to a spectrum and is directed back to the slit panel and recorded for further analysis on a photographic plate loaded inside a cassette. The high degree of dispersion gives the analyzed spectra a total width of 4–12 m from red to violet.

Fig. 17. Photograph of the solar spectrum (with dark Fraunhofer lines; horizontal strip in center) and a comparison spectrum of iron generated with an electric arc (with bright emission lines; upper and lower horizontal strips) taken in first order at a dispersion of $1\text{Å} = 0.75$ mm. (From Freundlich 1927b [22], by permission of Springer-Verlag)

The technical specifications of the tower telescope clearly copied the American model.[14] But the exteriors of the Potsdam and Mt. Wilson observatories could hardly have been more different. (Compare Figs. 10 and 24.)

Clearly, the relatively inexperienced architect had received this assignment only because no one knew at the time how much this particular building would contribute to architectural history. There was consequently no protest when Freundlich first suggested his personal acquaintance. But Mendelsohn used his chance to the fullest. In retrospect we see that it was this assignment that provided Mendelsohn's breakthrough, giving him access in the 1920's to well-paying clients and even winning him star status as an architect; at one time he was able to employ a staff of 40.[15] The full-page photo of the Einstein Tower on the front of the *Berliner Illustrierte Zeitung*, the world's first photo magazine, in 1921 is only one example of its extensive press coverage. (See Fig. 19; it was this photo, incidentally, that brought Mendelsohn together with the Jewish publisher Hans Lachmann-Mosse [1882–1941], son-in-law of Rudolf Mosse [1843–1920], founder of the well-known publishing house, and resulted in a commission to redesign the publisher's headquarters.)[16]

Mendelsohn had been provided with technical specifications only for the core of the tower building; he was given a free hand with the exterior. This was an unusual luxury, especially compared with commissions for office buildings, retail stores, or industrial complexes. His radical design did not pass unnoticed, however. The Building Surveyor's Office, which issued the building permits, noted that "the external design of the building is unusual but a little capricious, and the stylistic language does not in any way suit the other buildings of the Astrophysical Institute. However, as the latter originate from a time that no longer satisfies current tastes entirely and the new building will be erected at a slightly removed spot, there ought to be no objection to it."[17] Thanks to the runaway inflation rate, such reservations about the design were briskly swept aside by decree at the highest cabinet levels, so as not to delay the start of construction.[18]

The intervention of government officials in this project explains why the Einstein Tower could be erected without any cuts whatsoever, even though in the aftermath of the German defeat in World War I, new projects could usually be carried out only with the strictest economy.[19] After Germany's bitter loss and resulting political impotence and economic crisis, this monument to Einstein embodied the pride of the nation for its position in science.[20] As happened frequently in German history, glorification of genius compensated for a physical loss of power. When Bismarck and Wilhelm II had become historically obsolete, a monument was built to relativity theory and, ironically, to its anything but nationalistically minded inventor.

The location of this postwar project of prestige in Potsdam, on the outskirts of Berlin, is significant. At that time Berlin was, after Paris, the second art capital of Europe. Most of Germany's new talent was there, making major contributions to the development of modern architecture: Bruno and Max Taut, Mies van der Rohe, Walter Gropius, Hans Scharoun, Hugo Häring, Hans Poelzig, Artur Korn, and other progressive architects of "above-average ability."[21] Many were never able to develop their best designs beyond the drawing-board or plaster-model stage, and only very rarely could they realize their visions so freely as Mendelsohn was able to in this instance. His only constraints were the vertical dimensions of the tower and the horizontal extension of the laboratory (which were fixed by the focal length of the spectrograph and lenses), as well as the absence of windows on the south side of the building in all rooms directly adjoining the telescope's light path, to avoid light-scattering effects.

The assignment of the exterior design of a towerlike building was fortunate in another way as well. Towers had become a leitmotif of architecture in this period. The so-called Wedding Tower erected at the artists' colony at Mathildenhöhe in Darmstadt is an example. (See Fig. 20.) Mendelsohn and his colleagues produced numerous imaginative sketches with this theme, to express "dynamics, the impulse to gain height, and vertical ascent," as well as "the aspiration of higher and abstract things, of

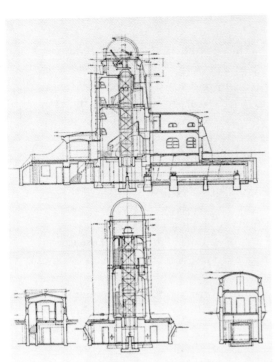

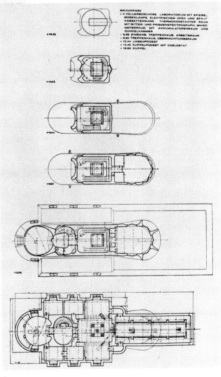

Fig. 18. Blueprint drawings of the Einstein Tower, vertical and horizontal cross sections. (From KB, by permission of Prof. B. Evers; cf. also Sharp 1966 [144–46], Limberg in Limberg and Staude 1994; and Eggers, ed., 1995 [93–95, 103, 105] for earlier versions)

coming closer to the elements, of rising above the throng, and the ecstatic in the literal sense."[22]

On comparing Mendelsohn's 1919 sketches or his plaster model (Figs. 14, 26) with the final construction, we can see how true to the original idea the final outcome was. Mendelsohn did have to abandon one thing, though: he had originally planned to have a spiral staircase encased in glass around the inner wooden tower, but the extreme sensitivity of the optics would not allow it.[23] He also had to accommodate a more stable foundation for the coelostat and lens in the dome, resulting in a stouter tower. But the design was not altered significantly. Mendelsohn wrote in a letter to Freundlich on May 12, 1920: "The 'tower' gets a slightly fatter belly, and all the proportions naturally change as a result. The arrangement remains the same. A guided tour through the Zeiss factory revealed a conformance between the technical bases and my optical vision. Proof of the primitivity of intuition."[24]

1 September
1 9 2 1
Nr. 36
30. Jahrgang

Berliner

Einzelpreis
des Heftes
75 Pfg.

Illuſtrirte Zeitung

Verlag Ullſtein, Berlin SW 68

Fig. 19. The Einstein Tower as featured on the cover of the *Berliner Illustrierte Zeitung*, September 4, 1921, shortly after completion of the building—in the foreground, building and painting equipment. (By permission of Ullstein Bilderdienst)

Fig. 20. Other tower designs of the time: *top left*, radio tower in Kootwyk, the Netherlands, by J. M. Luthman; *top right*, observation tower in Grenoble, France, by A. and G. Perret; *bottom left*, water tower in Nykøbing, Denmark, by E. Ambt; *bottom right*, the "Wedding Tower" in Darmstadt, by Joseph Maria Olbrich (ca. 1908). ("Wedding Tower" from Olbrich 1914 [pl. 1]; other photos © Ernst Wasmuth Verlag, Tübingen)

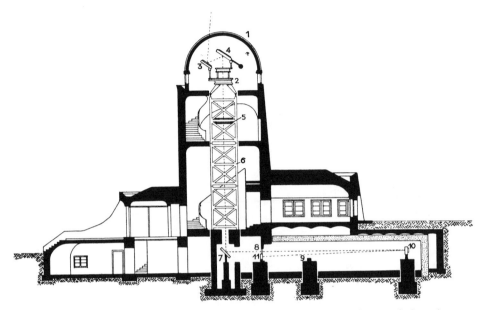

Fig. 21. Path of light inside the Einstein Tower in the original setup: below the dome (1) is the coelostat and mirror assembly (2–4), which directs the light (dotted line) through the wooden tower frame (6) to the redirecting mirror (7); from there the light is projected onto an adjustable slit (8) through which a desired section of light enters the thermally insulated underground laboratory for spectrographic analysis (with either the prism spectrograph [9] or the diffraction grating [10]) and photographing (11). (From Fürst 1926 [31]; cf. below, Figs. 22, 32–36)

The core of the building was predetermined, along with the building's scale; but its external shell, which alone sets the image of the building, was left entirely to the architect. The marked contrast between the building's function and its design is particularly apparent in Figure 21. The crux of the architectural problem lay in building a structure on a separately anchored wooden frame around the entire optical system. Wood was used instead of steel to help dampen vibration in the heavy instruments in the tower and thus prevent blurring of the images; and the separate foundation allowed the optics to remain unaffected by any movement of the building caused, for example, by gusts of wind against the external walls. The metaphor of core and shell could not have been better chosen.

Mendelsohn described what an architect usually had to tolerate:

In actual fact the architect is so tied to the purpose, materials, and construction, so dependent on the will of the person commissioning the building and on his available funds, that it is clearly reflected in the final work. . . . In our times—business, industry, commerce—demands are often made that make up 99 percent of the real impulse of the total work. But the addition of just 1 percent intuition transforms the "materials" into a masterpiece.[25]

In this instance, Mendelsohn was given the opportunity to add more than the usual 1 percent intuition. He was perhaps never again allowed such leeway in bringing his first inspired sketch, his architectonic idea, to full fruition.[26]

But what had defined this initial inspiration? What idea was being expressed in the first sketches and later developed? Surely it was more than simple self-expression, which had motivated many expressionists in the creative arts especially. Mendelsohn saw clear links between his conception and his knowledge of relativity theory:

Since learning that two previously strictly separate concepts—matter and energy—are only two different manifestations of the same basic material and that nothing in the universe is unconnected to the cosmos or unrelated to the whole, engineers have abandoned the theory of lifeless matter and have devoted themselves ever since to the deferential service of nature. They find consistent inter-

Fig. 22. Interior view of the Einstein Tower, looking upward through the wooden tower frame to the coelostat lens. (From KB, by permission of Prof. B. Evers; cf. above, Fig. 21)

relations in the most elementary phenomena, and their former arrogance gives way to an enthusiastic involvement in the creative process. The machine, previously a subordinate instrument to be inexhaustibly exploited, has become a constructive element within a new living organism.[27]

The Einstein Tower's "relaxed horizontal forms" balance out the "stresses of our fast-paced life"; it is an early incarnation of the search for holistic alternatives.[28]

Supported by allusions to the consequences of relativity theory, such as the mass-energy equivalency in the previous quote, many contemporaries justifiably saw the tower as an architectonic rendition of Einstein's theories. Whether this did justice to the theory is another question. In investigating the relationship between the layman and modern architecture in 1925,[29] Manning Robertson saw a glaring incongruity between the two, judging that Mendelsohn had grossly misinterpreted relativity theory. According to him the building was eccentric, bizarre and grotesque, a screaming example of incoherence:

It appears as a gaunt and curvilinear tower, with a base resembling a crazy concrete battleship, apparently independent of geometry and its derivatives. Its "purpose," we may take it, is to express the spirit of Einstein's theory. The expression may reflect the popular imagination of that theory, but it is not one that could be accepted by the world of science. . . . Instead of using a structure of revolutionary simplicity, the architect has designed a building of excessive complexity, not so difficult to understand as it must have been complicated to build. Hence, it exists, a travesty of Einstein's actual contribution and a monument to complication and bewilderment.[30]

A bizarre tower to express what many perceived as Einstein's equally bizarre theory of relativity? This conservative Fellow of the Royal Architectural Society does seem justified in inferring that Mendelsohn had probably not completely understood Einstein's theory. The Einstein Tower was erected at a time when all kinds of obscure speculations were being made about non-Euclidean geometry, the fourth dimension, and the meaning of Einstein's space-time; and the new worldview also left its mark in architecture.[31] One example, although of quite late date, is Siegfried Giedion's book *Space, Time and Architecture* (1941),[32] in which the author describes modern trends in the creative arts and architecture as a radical expression of a new conception of space and draws bold parallels between Apollinaire and Einstein, between futurists and Minkowski.[33]

Mendelsohn sent excerpts from this book to Einstein in November

1941 requesting his opinion of it. (See Fig. 23.) Fully aware of the practically unavoidable misunderstandings that had arisen from innumerable
misinterpretations of relativity by philosophers and journalists alike,[34]
Einstein did not think much of unrelated fields' borrowing the vocabulary
or isolated concepts from his theory. His humorous response to Mendelsohn, in the form of a quatrain, was:

> Nicht schwer ist's Neues auszusagen,
> wenn jeden Blödsinn man will wagen.
> Doch selt'ner füget sich dabei,
> dass Neues auch Vernünftig sei.

> (It's not hard to make new predictions if you're willing to venture any
> old nonsense; but it's rare when the new idea also makes sense.)

His handwritten postscript to this typed note was explicit: "It's nothing
but clever bullshit, without the least rational basis!"[35] Mendelsohn also
distanced himself from Giedion's interpretation, as is illustrated in the
draft of a letter to Einstein in which he points to the "great confusion"
Giedion creates among young architects by mixing up in his analogies the
concepts of four-dimensional space-time (in Einstein's relativity theory),
three-dimensional space (in architecture), and two-dimensional planes (in
painting).[36]

Through his friendship with Freundlich, who was not only completely versed in the secrets of the theory but also was one of its first and
most qualified publicists, Mendelsohn was certainly well placed to inform
himself about relativity; and Mendelsohn apparently took full advantage
of this in the period between 1917 and 1920. Thus the architect applied
his own understanding of the union of space and time dictated by special
relativity, and of the general theory's non-Euclidean space-time structure
and closed yet infinite cosmological field equation solutions. In any case
the result was no austere functional building like the tower telescope at the
Mt. Wilson Solar Observatory, but a monument to relativity theory,[37] an
allegory of the temporality of space that Einstein had recognized. The
architectural historian Bruno Zevi comments on the Einstein Tower:
"We are far away from cubism here, from a 'fourth dimension' which
refused to recognize the thickness of building materials. Here the time
factor was intrinsic to the autogenesis of the building, making it an earthly
event, passionately conquering its space."[38] The Bauhaus movement's
"functional construction" was not Mendelsohn's intent here; he aimed at
"dynamic construction." The function of the building, although the

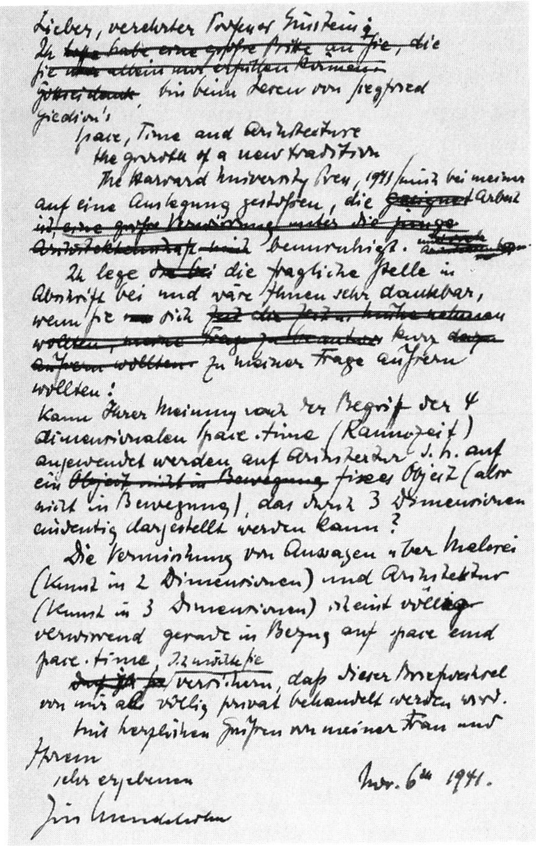

Fig. 23. Draft of a letter of Mendelsohn's to Einstein, November 6, 1941. (From KB, Einstein–Mendelsohn correspondence folder, by permission of Prof. B. Evers)

source of the inspiration, nevertheless does not dictate its essential form. On the relationship between function and form Mendelsohn states:

From the outset we architects must subject our planning *as a matter of course* to the material requirements and constructive relationships; we must simply regard them as *preconditions* to the entire organization of a building—but we must know that they are only one component of the productive process. But despite large dimensions and a clear implementation of the technical resources, this alone does not make architecture.[39]

The Einstein Tower's flowing, well-rounded forms place it conspicuously outside the building conventions of early postwar Germany. Contemporary viewers regarded its design as a symbol of the radically new, exotic, and "modern"—like relativity theory itself, which, though not considered such by Einstein, was nevertheless seen by the general public as a radical break with theoretical tradition. We find the same pattern of reception for both the observatory's design and its main object of research: enthusiasm by avant-garde critics and rejection by conservatives. Philipp Frank, Einstein's successor in Prague and biographer, describes the contemporary "average citizen's" impression of the Einstein Tower:

The institute had the shape of a tower and was built in a modern Berlin style that the average citizen took for something between a New York skyscraper and an Egyptian pyramid. The tower was generally known as the Einstein Tower, and its design alone annoyed nationalistic circles, which at that time preferred a style that rather followed German medieval models or at the very least classical antiquity.[40]

The Potsdam Telegraphenberg is an excellent site for viewing the contrast between postwar architecture and the retrospective style of the turn of the century. The Einstein Tower, with its uniformly whitewashed, smooth, cement facade, stands right next to the brick-faced Astrophysical Observatory, built in 1890 and ornamented in the Moorish style. (Compare Fig. 7, above, and below, Figs. 24, 25.) Mendelsohn's unadorned, stylized architectural sculpture of 1920 contrasts sharply with the massive, meticulously ornamented refractor dome of 1899–1900 and the main building dating from 1876–79.[41]

The new building became a tourist attraction, drawing crowds of visitors to the observatory grounds. The astronomer Harald von Klüber was often enlisted to conduct tours, and he later reported to Erich Mendelsohn's widow about his experiences as a guide:

I soon had to discover that to my slight disappointment the interest of the visitors was not so much captured by the interest for astronomy or as a matter of fact for Einstein's famous theory, as I had hoped for. The visitors wanted to see the new interesting and controversial building and its famous architecture. Eventually, and in order to make the best of the situation, I learned to turn my usual explanatory talk from modern solar physics to one of comparative architecture. I used to point out with hopeful emphasis that much of the surrounding buildings were constructed and styled at the end of the last century in the cool, sober Prussian style with the red bricks of the Mark Brandenburg. They well fitted to the classical Euclidean concept of mathematics and to the atomic structure of matter, consisting of discrete units, as one understood it at the time of the outgoing century. Whereas the new style of Mendelsohn reflected very well the aspects of quite modern technology, mathematics and physics with their complicated, nevertheless highly aesthetic and attractive ideas, with transcendent functions, complicated swinging forms and elegant curvatures. This contrasted to the much simpler, orthogonal Euclidean system of older days and it reflected in some way the just now starting breakdown of the simpler atomic models in favor of the new more transcendental conception of a smooth changeover between matter and energy as first pointed out by Einstein.[42]

Von Klüber himself was at the time not altogether convinced of the bold parallels he was drawing for his visitors in an attempt to awaken their interest in his field, but as we learn in the same letter, later on he had notably fewer reservations in this regard:

By myself I thought this all to be probably a fair nonsense and that the architect might in reality never have contemplated his creation under such parallels. . . .

Today, after decades, and after having seen myself so many more different architectures all over the world, I am no more so sure that my comments at that time were quite so wrong as I thought them to be. Certainly, an artist or a first-class architect, as Mr. Mendelsohn was, does not need to think or even know much and in detail about rationalistic relationships or justifications. He probably feels and experiences the idea and the style and the trend of his time on an even better and certainly more sensitive level than rationalistic analysis could do.[43]

Some architectural historians regard the Einstein Tower as the principal work of architectonic expressionism[44] and include among their arguments in support of this stylistic classification Mendelsohn's close personal contacts with expressionist painters (the "Blue Riders") during his Munich period, as well as later with the Dutch expressionist generation. Mendelsohn conceded this himself in 1920 in a special issue of the Dutch publication *Wendingen*, which ran a feature article on his work.[45] But we must clarify what is meant by "expressionism" in architecture. The historian Jürgen Joedicke, for instance, provides a definition that emphasizes its link to painting:

Expressionistic architecture was an attempt to translate the striking physiognomy of contemporaneous surface art to the three-dimensional architectonic structure. The rudely jutting profile and the plastic, animated structural mass lent an expressive note to the edifice. Light, above all, both natural and artificial, was incorporated effectively.[46]

These criteria certainly apply to the Einstein Tower, just as does Wolfgang Pehnt's definition of the formal characteristics of expressionism in his classic study *Expressionist Architecture*:

Fig. 24. Side and rear views of the Einstein Tower. (From KB, by permission of Prof. B. Evers)

"a spatially plastic structure shaped so that it can be experienced in all
    dimensions,"[47]
a plastic treatment of building materials, so that they appear to be
    "molded, cut into, hollowed and gouged out,"
and not least, uniqueness of design, in sharp contrast to the Bauhaus
    search for rational, mass-producible shapes suitable for public housing.

Others disputed this stylistic designation, however, basing their views
on the essential differences between Mendelsohn's "harmonious" designs and the "disharmonious, often screamingly dissonant" expressionist
works in the other arts, especially painting.[48] The architectural historian
Kathleen James also emphasizes that despite close personal links with
several expressionist artists Mendelsohn always "maintained a personal
and theoretical distance from his fellow expressionist architects," none of
whom he had apparently known before the war.[49] Mendelsohn never
considered himself an expressionist. Throughout his life he resisted the
self-expression evident in the work of Kokoschka, Nolde, or Barlach,
suggesting instead an effort to express something about the nature or
essence of a structure itself.[50] He considered himself a disciple of Henry
van de Velde during this period, picking up where the Belgian architect
had left off.[51] This labeling of the Einstein Tower telescope seems rather
confusing. It would be preferable if Mendelsohn's architectural style,
which is closely related to Rudolf Steiner's in Malsch and especially in
Dornach (ca. 1911–24)[52] and Antoni Gaudi's in Catalonia (after 1900),
were to acquire some new name—"utopian" or "fantastic architecture,"[53]
or perhaps more appropriately "dynamism," a term Mendelsohn himself
used occasionally when referring to the Einstein Tower's relative architectonic tensions and general forward motion.[54] However, Mendelsohn's
approach to dynamics and form also differed from the *dinamismo architet-
tonico* demanded in the "manifesto of futurist architecture." We must thus
also disagree with David Eddy's suggestion that the Einstein Tower should
be regarded as "the only authentic Futurist building."[55] A glance at the
layout of the tower (Fig. 18) reveals the perfect symmetry of the building
along its main axis as well as the near symmetry of the secondary vertical
axis, in complete contrast to the asymmetric designs in the Amsterdam
architecture of this period.[56]

There is an indisputably organic element to the tower's shape. This
concept is present throughout the entire history of the reception of the
tower, starting on the day that Mendelsohn guided Einstein through the

newly completed building. Einstein reportedly listened with interest to Mendelsohn's explanations without saying anything. During a meeting of the Einstein Foundation board immediately after the tour, however, he suddenly turned to Mendelsohn and whispered in his ear the single word "organic."[57]

The structure certainly does seem to have been poured in a single casting, in a single creative impulse. With its positive and negative forms, its horizontal and vertical lines, the building was composed with an eye to the whole. Mendelsohn himself would use organic metaphors in speaking of a "monolithic" treatment of the entire structural complex: "We must obviously admit that the Einstein Tower is clearly an architectonic organism. Leave aside its reasons for not being simultaneously a purely functional organism. But it seems to me that you cannot remove one part of it, neither its mass, its motion, nor its logical conclusion, without destroying the whole."[58] Every detail is harmonized within the whole, from the gutters, with their semicircular projections subdividing the convex curvature of the outer wall,[59] to the sweep of the walled entrance stairway (cf. Fig. 25), which assumes the rhythm of the facade design and repeats the tower shape in miniature. Many commentators have used metaphors of rhythmicity and motion to describe the inherent dynamics of the building. Adolf Behne, for instance, described its architecture as "working intensely with movement and anthropomorphism." He asserts: "The entrance 'sucks in,' the walls 'guide,' the stairway steps 'sweep,' etc."[60] Kathleen James even coined the term "dynamic functionalism" to describe Mendelsohn's style.[61] Paul Ferdinand Schmidt's interpretation was that Mendelsohn had allowed "the space within the spheres of musical emotion to take shape in plastic, nonarchitectonic rhythms and erected a monolithic tower of, as it were, molded concrete mass . . . a last tribute to the agitated and chaotic revolutionary times, known in art as expressionism, a direct transferral of the dynamic principle in crystalline formation, without relaying architectonic methods."[62]

The tower's exterior concrete casing gives it a taut, skinlike appearance accentuated by cleverly contrasted lit and shaded areas on its curved surfaces. One notes, for example, the shadows of the curiously sculptured gutter spouts against the expansive free surfaces, or the conic-section shadows in the curved window spaces. (See Fig. 25.)

The cylindrical tower section, which is strictly subdivided horizontally by the forward-shifted window arches, automatically calls to mind Karl Blossfeldt's detailed photographic plant study *Art Forms in Nature*.[63]

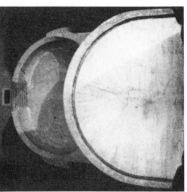

Fig. 25. Detail photographs of the Einstein Tower: *top left*, profile of the gutter spouts; *top right*, aerial view of the entrance area; *bottom left*, entrance stairway; *bottom right*, close-up of the concrete wall of the entrance stairway. (From KB, by permission of Prof. B. Evers)

On the other hand, the periodicity of the 16-meter-tall tower also contains elements of the futuristic machine aesthetics prevalent at the beginning of this century: the window arches' strict regularity, or the streamlined edge into which the smooth southerly front of the tower tapers off.

From the outside, the building looks like a monumentally enlarged clay sculpture. In his 1927 handbook of modern architecture Gustav Adolf Platz called the Einstein Tower "the sculptor's youthful, carefree audacity," and an "(almost molded) structure poured of soft concrete."[64] Many others have emphasized the tower's appearance of having been "molded in the hands of a creator demiurge."[65] An examination of the model completed shortly before the beginning of construction strengthens this impression. A contemporary wittily linked the building's compactness with its unofficial namesake: "Ein Stein" (One Stone),[66] underscoring the general perception of the tower as a monument to Albert Einstein,[67] who had become world-renowned since the spectacular light deflection measurements of Eddington and Crommelin. The originality and monumentality of Mendelsohn's design was not always appreciated. One of his sharpest critics, Paul Westheim, wrote in 1923:

There the approach was not from the functional point of view of an engineer, but was still rather an awkward flirtation with engineering and a monumentality that despite the charming modern form of expression is essentially akin to the Battle of the Nations Memorial and the Bismarck on the Rhine. This tower near Potsdam is a gigantic poster, an advertisement, less of the observatory than of the originality of the builder. It is a Mendelsohn Tower. It is not actual architecture that we have on display here, but architecture and the architect's personality.[68]

As Manning Robertson pointed out in his analogy of a "crazy concrete battleship," particularly when standing inside the workroom you instinctively feel as if you were on the bridge of a ship.[69] Shiplike elements—rows of porthole windows, elongated hull construction—transform the Einstein Tower into a "spaceship" in which the modern astrophysicist travels through the universe of Einstein's space-time. In an interesting parallel the tower reminded Kurt Joël "a little of a submarine periscope."[70] Like a periscope, the Einstein Tower feeds light into its main area, which is also almost completely submerged, while attempting to shut out all other stimuli from the outside world. "Constant, precisely regulated heating and ventilation, the possibility of shutting out both light and air, space minimization not only in the floor plan but also in room height, central power supply—all these are characteristic of rooms

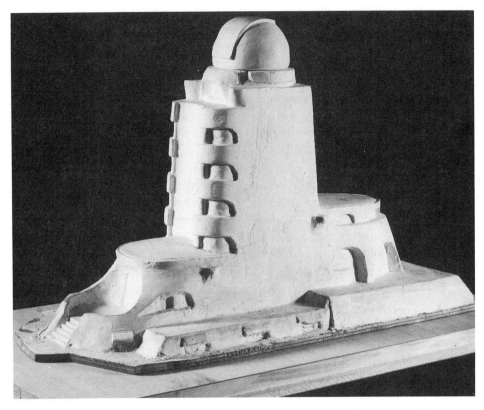

Fig. 26. Mendelsohn's plaster model of the Einstein Tower, of about 1920. (Photograph by courtesy of Mrs. Ursula Seitz-Gray, reprinted with permission of the German Architectural Museum, Frankfurt, where the model is located)

first realized in the steerable artificial environments of submarines and airships."[71]

Functionality took second place to appearance in the conception of the building. Researchers presently engaged at the tower report that the workroom is lit adequately by sunlight only in the springtime, before the surrounding trees are in full leaf, because the windows are set so deeply into the facade. In 1927, when the tower was extensively renovated, the horizontal external surfaces such as windowsills and gutters were lined with sheet metal, further accentuating the horizontal elements of the building.[72] The austere gray furniture of the workroom is planned down to the last detail (including the lamps, radiators, shelving, etc.), but is rather uncomfortable to use over long periods. Fortunately, most of the work takes place in the underground laboratory. Hein Köster writes on the interior decoration:

With the interior decoration Mendelsohn imposes an ascetic aristocratic science pathos: furniture, lighting fixtures, and coloration are extremely frugal, subordinating themselves entirely to the demands of intense work. No luxury. No diversions. A temple of science.

The plastic undulations and shapes of the building are taken up in the furniture, yet sobered by the exclusive use of straight or angled lines.[73]

The stark contrast between the rounded forms of the exterior and the strictly geometric composition of the workroom is stylistically striking. The architectural historian G. K. König explains this as the result of the time that elapsed between the planning and completion of the shell of the building, in 1919–21, and its interior decoration later, when Mendelsohn was turning to a new functionalism inspired particularly by Bauhaus experimentation with industrial design.[74] The similarity of the interior, such as the entrance stairway, to film sets in *Dr. Caligari* (1920) or *Metropolis* and other Universum-Film-A.G. (UFA) productions from 1924 onward is also frequently pointed out.[75]

Like other avant-garde interiors, this stylistic "reunification of all artistically creative disciplines," such as Walter Gropius called for in 1919, placed far too many demands on the human being. The impractical heating units and lamps were replaced and the overnight room was completely redecorated long ago. It is only because the astrophysicists feel a sort of curatorial respect for the memorial value of their workplace that some of the workroom's original furnishings have survived to this day. But we are already familiar with the subordination of inhabitant to aesthetics in Bauhaus architectural designs or in the Art Nouveau "synthesis of the arts" (*Gesamtkunstwerk*) at the Darmstadt Mathildenhöhe, where sometimes even the residents' attire was prescribed.

Rather than attempt to present Mendelsohn's programmatics, let us look at his own description of the revolutionary spirit affecting young intellectuals after the political and economic disintegration of 1918:[76]

The simultaneous arrival of revolutionary political events and fundamental changes of human interaction in commerce and science, in religion and art, automatically provides a belief in new form—law and control—which evidences the birth of a new lawfulness brought on by catastrophes in the history of the world.

When forms shatter, they are only giving way to new ones that have long been there but only now come to light.

The reshuffling of the spirit of the times means, in architecture's particular circumstances, new tasks from the changed building purposes in transportation, commerce, and culture; new construction options from new building materials: glass—iron—concrete. . . . We are at the very beginning, but already can choose among its paths of development. . . . What actually happens has value *only* if it emerges in the flush of a vision.[77]

The new options for construction that iron and concrete presented especially appealed to Mendelsohn; they had the potential to surpass the limits of static masonry construction, which remained rigid, angular, and ponderous. He sought dynamism and active tension (*Spannungsbewegtheit*) in his space studies (*Raumstudien*).

Mendelsohn was aware that such considerations were revolutionary attempts to "overturn . . . archaistic constraints." He also knew full well that it still remained for him to provide "proof of the constructive materializability" of his "visions":

The adjustment of our static senses to the tension between iron and cement in place of the previously held principle of support and load has to be made gradually, through a long sequence of approximations. . . .

Only with the awareness that concrete is a filler building material, only with the linkage of both construction elements—concrete takes on the shearing stress, iron the tensile stress—does iron reject its hybrid and exclusively technical character, thus achieving the uniformity of a surface, the spatiality of a mass; only now does it gain the potential of presenting a challenge to new form.[78]

The Einstein Tower commission was Mendelsohn's chance to realize his vision, but technical difficulties arose in carrying out his original plan of constructing the tower exclusively of steel-reinforced concrete. Photos taken during the construction show that the main walls of the tower, with the exception of the uppermost section under the dome, the entrance area, and parts of the workroom segment to the south, were constructed of conventional brick masonry and were only finished with a thin cement facade. (Cf. Figs. 25, 26, 29.) There are two different explanations for this in the older literature:[79] according to Mendelsohn's widow, cement was being rationed during the construction phase, and there was not enough available for the entire building; other sources report, however, that the builders felt incapable of carrying out the complicated formwork necessary to produce the exterior's many curved surfaces. The second version

Fig. 27. Original furnishings of the Einstein Tower: *above*, first-floor recreation room; *below*, workroom on the ground level. (From KB, by permission of Prof. B. Evers)

Fig. 28. The Jena planetarium under construction, 1923. (From Bauersfeld 1925)

seems more plausible, not least because one of the cement suppliers was
otherwise involved in the tower's construction. The Dyckerhoff Company also supplied the "Torfoleum" thermal insulation for the measurement room and probably would not have stinted with cement, or would
even have economized on other projects if necessary, because of the
prestige attached to its participation in this one. In addition, because the
outlay for building materials was minimal in comparison with overall
expenses, it would be difficult to argue that cost was the limiting factor.
On the other hand, building crews still had little experience with formwork and cement casting techniques, particularly with rounded forms.[80]
Poured-concrete construction was employed successfully only in 1923,
when the Carl Zeiss and Dyckerhoff firms built the Jena planetarium.[81]
(Cf. Fig. 28.)

According to Jörg Limberg, the Potsdam curator at the Office for the
Preservation of Monuments, neither of these versions is accurate. He
asserts that it was primarily for financial reasons that Mendelsohn finally
opted for a combination of old and new building techniques to minimize

Fig. 29. The Einstein Tower under construction ca. 1921, with the brickwork clearly visible in the middle. (From KB, by permission of Prof. B. Evers)

Fig. 30. The Einstein Tower at the end of World War II, after damage in an air raid on April 14, 1945. (From KB, by permission of Prof. B. Evers)

actual construction costs, underestimating the long-term problems this would eventually cause.[82] The reinforced concrete segments at the entrance and top of the building constitute proof that the technical problems involved in producing these curved surfaces could be mastered, but only with a considerable investment of labor. According to Julius Posener, Mendelsohn later exclaimed: "Dear me, never again! We had to employ shipbuilders to do the formwork."[83] The shipbuilding experience Freundlich had garnered during his brief practical training at the Vulkan wharf (see Chap. 1) must have been of some use during this stage of the construction.

The Einstein Tower survived World War II, although it was damaged in an air raid on April 14, 1945, which destroyed a building nearby. (See Fig. 30.) The instruments remained virtually unscathed, however, and the tower was quickly repaired: observations were resumed in the fall of that year. By 1950 renovations were complete. The tower telescope, equipped with more modern instruments, is still in use today for solar spectrum measurements, particularly measurements of the magnetic fields in and around sunspots, as well as analyses of their structure and dynamics.[84]

CHAPTER 8

Research at the Einstein Tower During the Freundlich Era

Unlike the main structure of the Einstein Tower, which was completed in 1921, the optics required for the astronomical investigations (the entire erect telescope, including the coelostat in the dome and the grating and prism spectrographs) was installed only at the end of 1924. With the inaugural committee meeting of December 6, 1924, the tower was officially in operation. According to the progress reports to the Einstein Donation Fund from the years 1920–23, Freundlich continued the microphotometric analyses he had begun in 1919. This research was of importance to the Bonn physicists Leonhard Grebe (1883–1967) and Albert Bachem (1888–1957), for Freundlich placed the Koch microphotometer at their disposal in 1920.[1] He also used an electrical resistance oven that the Badische Anilin- und Sodafabrik (BASF) chemical company had provided on the instruction of its director, Carl Bosch,[2] using it in conjunction with a scientist from the company's laboratories at Ludwigshafen. Commuting there initially until a spectroscopic resistance furnace had been installed in his own laboratory,[3] Freundlich worked together with Dr. Ernst Hochheim[4] (1876–?), in particular on the cyanogen band spectrum.[5] Because it was pressure-independent, this band had an important place in the discussions on gravitational redshift, and its origins were still controversial.[6]

Freundlich and Hochheim also investigated the appearance of spectral lines for various elements at different temperatures and the influence of carbon on the cyanogen band spectrum in particular. Once the electrical resistance oven was in place, it was possible to generate the spectra of

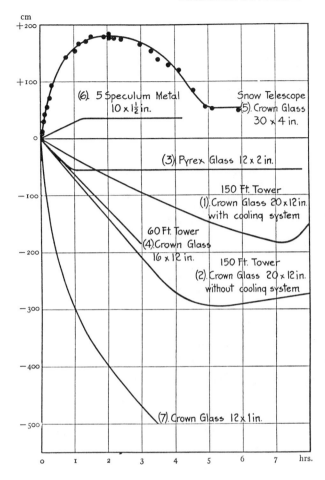

Fig. 31. Expansion characteristics of Pyrex glass in comparison with other materials: effect of time of exposure to sunlight on the focal length of coelostat and mirror systems. (From Pettit 1923 [210], by permission of *Astrophysical Journal*, University of Chicago Press)

glowing substances while avoiding the electrical discharge commonly produced in previous laboratory spectra. Temperatures up to 3,000° C and pressures between 1/1,000 and 1/10,000 atmosphere could be obtained to simulate physical conditions on the surface of cooler stars.[7]

Pyrex glass was chosen for the entire lens and mirror system. This glass was a major improvement over the types in conventional use. As Figure 31 illustrates, even after many hours of exposure to sunlight, it

Fig. 32. A small, flat redirection mirror, tilted at an angle of approximately 45 degrees, guided the incident light from the Einstein Tower's shaft to a slit panel in the underground laboratory. (From Freundlich 1927b [29], by permission of Springer-Verlag)

hardly expands at all. The resulting images are less distorted and precision measurements are thus more reliable.

After the optics had been installed in the winter of 1924, time was still needed to adjust the gratings, which, as Freundlich reported, necessitated construction modifications right at the outset.[8] The new technology furnished by the Carl Zeiss Company in Jena presented a particular challenge to unversed experimenters. Although the basic idea of the tower telescope had been borrowed from Mt. Wilson, precise technical details like mirror and lens sizes, measurement instruments, laboratory light sources, and so forth, still had to be worked out. Planning of the optics began in 1921, after the main construction had been completed: "Construction of the new institute in Potsdam is close to completion. The design of the installations, particularly the new telescope, occupies all my time at the moment, for we are building a new type and I must assume responsibility for everything—that means, essentially, considering every last detail twice and three times over."[9] The optical firm's staff also had to adjust to the unfamiliar specifications of this new type of telescope. This explains in part the tardy arrival of the optics. (The Zeiss Company expected similar orders of coelostat optics in the future, however: one example is the Göttingen solar tower, built in 1943; others are the solar towers in Tokyo and in Arcetri, near Florence.)[10] The original instrumentation (which is no longer in use) is shown in Figures 34–36.)

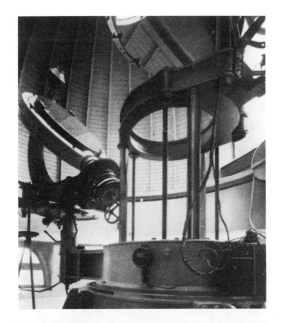

Fig. 33. *Above*, Coelostat and main mirror in the dome of the Einstein Tower, which are adjusted to project sunlight or starlight downward. (From Freundlich 1928d) *Below*, Zeiss Company advertisement featuring the equipment it produced for the Einstein Tower. (From Villiger 1926: 6)

Fig. 34. *Above*, View into the basement laboratory of the Einstein Tower as originally equipped. The tilted mirror that redirected the incident light and the small aiming telescope were located on a pedestal in the center between the four pillars of the tower. To the right, an electric arc; to the left, the electrical resistance furnace that projected the light it produced through a series of lenses to the slit panel in the center of the room (behind the redirection mirror). *Below*, View of the electrical resistance oven when open. (From Freundlich 1927b [17], by permission of Springer-Verlag)

Fig. 35. *Above*, The Einstein Tower's spectrograph with a flat grating of 12.5 cm × 9 cm, 20,000 lines per inch. The camera's collimator lens had a focal length of 12 m and slid along a carriage for focusing. The focusing motor, which was also used to rotate the grating, was controlled from the slit panel. *Below*, Side view of the prism spectrograph. Light traveled a distance of 3 m from the slit to the lens, which focused it into a parallel beam. It then passed through the three prisms; the silver-plated rear face of the third prism reflected the light through all the prisms a second time, and the resulting spectrum was then transmitted through the same lens to the film cartridge for registration. The entire prism table was tilted slightly, so that the returning light fell below the slit onto the cartridge. The third prism (with the silver-plated rear face) could be adjusted from inside the laboratory to allow a selected color region of the spectrum to pass through the lens. The mechanism for turning the prism can be seen on the extreme left edge of the prism table. A 12-volt motor adjusted the lens position along a parallel carriage for focusing. (From Freundlich 1927b [23, 26], by permission of Springer-Verlag)

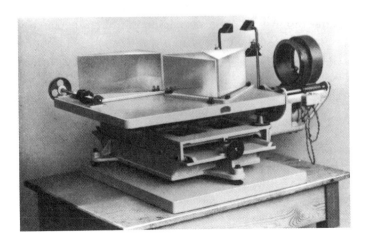

Fig. 36. H. A. Brück in the Einstein Tower's underground laboratory ca. 1930, before the slit panel onto which the solar image was projected, on the wall separating the workroom from the darkroom. Fine-adjustment controls were situated close to the seated observer. (By permission of H. A. Brück)

During the installation phase one of Freundlich's assistants, Felix Stumpf (1885–?),[11] had such difficulty with the instrumentation that he came to the conclusion that the telescope's design was faulty. When he reported this to his superior in 1922 upon the latter's return from a trip abroad, Freundlich wanted to fire him on the spot.[12] Stumpf turned in protest to Einstein, who solicited an opinion from James Franck, a corresponding member of the Einstein Foundation board. (See Chap. 10.) This opinion gives us an idea of the difficulties German astrophysicists faced in setting the tower telescope in operation, lacking experience with such apparatus:

As I understand it, Mr. Stumpf is faced with a problem that he could not solve, but that another younger physicist likewise would be unable to solve. For it is clear that the most qualified spectroscopist with many years of experience in this field might be just about good enough to tackle the immense task. We must consider that subtleties that are at the limit of feasible observation must be determined here; moreover, with these observations we are attempting to compete against a man like St. John. Naturally Freundlich cannot do this either—not that I want to underestimate in any way his abilities as a theoretical astronomer.[13]

Freundlich was forced to familiarize himself with the tower telescope's optical system. (For details of this system, see above, Chap. 7 and Fig. 21, and below, Figs. 32–36.) He managed within a few weeks.[14] Such technical difficulties are fairly typical of national pilot projects. Once they had been surmounted, the subsequent work plan was as follows. During the day, sunlight spectra were compared against spectra generated in the electrical resistance oven, using the grating spectrograph. The ultimate goal was to test the relativistic prediction of gravitational redshift as well as recent theories about the solar atmosphere. At night other observers analyzed light from bright stars using the translucent prism spectrograph.

Some of the findings from the tower telescope research or the instruments in the attached laboratory during Freundlich's directorship deserve particular mention, as well as findings by the theorist group. The Russian-born Emmanuel von der Pahlen (1882–1952)[15] redetermined the dependence of the sun's rotation rate on the solar layering in 1925–26, and in 1928 he collaborated with Freundlich in an attempt to find a dynamic interpretation of the so-called K effect through stellar analysis.[16] Harald von Klüber (1901–78),[17] and from 1928 on Hermann Alexander Brück (b. 1905),[18] conducted microphotometric analyses on absorption lines in the solar spectrum. In the course of these investigations a collaboration was established with the research team headed by Marcel Gilles Jozef Minnaert and Jakob Houtgast in Utrecht.[19] The physicist Karl Wurm (1899–1975)[20] introduced quantitative methods in the astrophysical analysis of molecular band spectra, after 1933 concentrating on comets and interstellar matter.

From the end of 1928 Freundlich's time was taken up with the expedition to Takingeun, Sumatra, to observe the total eclipse of May 9, 1929. (See Chap. 9.) The Potsdam expedition's solar corona observations were far less controversial than Freundlich's own light deflection results. These spectroscopic and photometric analyses were carried out principally by Walter Grotrian (1890–1954) and Harald von Klüber.[21] Grotrian's research on the sun's corona in particular gained international recognition. His analysis of the corona spectrum revealed that some spectral lines could not be related to terrestrial emission lines. He and Bengt Edlén soon determined that their presence was due to highly ionized atoms.[22]

Direct tests of gravitational redshift are conspicuously missing from this survey of successful research projects at the Einstein Tower. The Einstein Institute's annual reports note that the Potsdam observers did not yet feel prepared to tackle this principal task of the Einstein Tower:

The research aimed at testing relativity theory has not yet gone beyond the preparatory investigation stage. A larger research program with the same goal that very noteworthy physicists in America recently carried out unsuccessfully has shown again that it is useless to attempt to approach the question of the shift of spectral lines in the solar spectrum against the same lines from a terrestrial light source without comprehensive and systematic preliminary examinations. We have therefore tried primarily to create a terrestrial light source that promises to be a suitable comparison light source; and we have started a particularly broad analysis of the structure of spectral lines in the solar spectrum. . . . Only when we have gained complete insight into all relevant factors will we set out to actually test the theory of relativity.[23]

Research in this area, however, particularly at Mt. Wilson, was advancing rapidly. After initial negative results, in 1923 Charles St. John became convinced that gravitational redshift was verifiable in the solar spectrum, and his colleagues soon agreed with his abrupt change of conviction.[24] Faced with the complex problems of solar physics, the Potsdam astronomers were not yet on a par with their American counterparts, who by this time had decades of experience with the instruments behind them; and the rapid advances in the field threatened to push the German laboratory to the side. Whereas astronomers in the United States were eventually able to draw definite conclusions about gravitational redshift in the solar Fraunhofer spectrum around 1923, Potsdam had not yet even begun to make the necessary measurement series.

In addition, financial contributors had to be placated, initially with the argument that such a "preparatory and orientational observation stage" was necessary.[25] These included the above-mentioned development of suitable light sources for use in the comparison of solar and terrestrial spectral lines.[26] The need to catch up with American astrophysical research is emphasized quite dramatically in Freundlich's progress report of 1926, which attempts to excuse the institute's lack of presentable results on relativity:

The large scientific field of solar research has been studied intensively in America with ample funding for over 25 years. For lack of lasting contact with this existing experience and tradition, it is not possible for us to tackle the problems immediately without first having collected a certain amount of experience and knowledge. At the moment our new institute is certainly the only European institute with the means to conduct solar research again on a completely equal level with the large American observatories in California. But naturally it will take a few years before the institute's members have become completely familiar with operating the tower telescope's large and complicated instruments and before the degree of observation accuracy reaches the point required for really substantial advances in the existing questions.[27]

Freundlich attempted to stimulate more cooperation between Potsdam and Pasadena, in hopes of attaining the same level of competence more rapidly. In an application to the research foundation Notgemeinschaft der Deutschen Wissenschaft in 1925 for a grant to travel to Mt. Wilson, Freundlich writes:

Ever since the erection of the tower telescope in Potsdam, Germany also has had the opportunity to take part in modern solar research, which developed in the last decade with the aid of the tower telescope at the Mt. Wilson Solar Observatory. In particular, testing of the theory of relativity can now commence at the new Potsdam institute with means that are equal if not superior to the American observatory's. At the same time the methods of measurement employed at the Potsdam telescope for the precise comparison of wavelengths must catch up with American methods.[28]

This direct competition with the Mt. Wilson Solar Observatory had been self-imposed, however, through the choice of analogous apparatus; and it features in all of Freundlich's petitions, submissions, and reports to institutions that promoted the development of science in this period. But the strategy worked: his expedition's application was granted. Freundlich traveled to Sumatra for the solar eclipse and then to Mt. Wilson, where he stayed for a number of weeks.[29] Upon his return, Freundlich announced in his annual report to the Einstein Foundation "comparison measurements against related observations at Mt. Wilson."[30] But this plan never materialized, and no substantial collaboration arose between the two observatories.[31]

On January 1, 1927, Freundlich became a member of the Physics Committee of the Notgemeinschaft der Deutschen Wissenschaft.[32] In this capacity he joined his colleagues Eberhard, Münch, Wilsing, Biehl, Birk, Hassenstein, and von der Pahlen from the Astrophysical Observatory as well as Courvoisier, Struve, Goldstein, and Prager from the observatory at Babelsberg. Freundlich's newfound influence probably explains the conspicuous increase in the number of successful applications to the Notgemeinschaft for financial support and equipment from 1927 onwards

TABLE 5
Summary of Freundlich's financial support in Berlin, 1914–33

Date	Purpose	Source	Amount[a] (marks)	Type
1914	Crimean expedition (re. light deflection)	PAS	2,000	l
		E. Fischer	3,000	l
		G. Krupp	3,000	l
1918–20	Research contract to test GTR	KWIP	6,000	a
Feb. 1919	Microphotometer (Toepfer)	KWIP	500	l
July 1919	Grant for electrometer	KWIP	300	l
Dec. 1919	Cost-of-living salary increase	KWIP	300	l
	Cost-of-living salary increase (for 1920)		2,000	l
1920–21	Einstein Donation Fund for Einstein Institute	Private	1,190,000	l
	Prussian state government subsidy	MC	200,000	l
1921	Operating costs grant	MC	20,000	a
1921–	Donations for operating costs	Private	6,000	a
1921	Travel grant (Jena, Ludwigshafen, England)	MC	3,000	l
1922	Operating costs grant	MC	30,000	a
1923	Verification of astrophysical consequences of GTR	NG	–?–	l
1924	Grant for installation work (tower telescope)	MC	7,000	l
1925	Mechanic and half-day laboratory assistant salaries	NG	3,000	l
	Travel grant (Mt. Wilson)	MC	4,500	l
1926	Research grant	MC	4,000	a
1926–	Staff salaries grant	MC	9,000	a
1927	Modification of a plate-measuring instrument for evaluation of 1926 solar eclipse expedition	NG	900	l
1927–33	Scholarship: H. Brück	NG	150	m
	Scholarship: A. Unsöld	NG	250	m
Feb. 1928	Grant for instruments (additional components to new BASF vacuum oven)	NG	7,720	l
Aug. 1928	Grant for precise determination of line profiles and their fine structure (e.g., pumps, monochromator, Fabry-Pérot interferometer)	NG	1,540	l
Dec. 1928	Entire photographic equipment for eclipse expedition	C. Bosch	ca. 3,000	l
May 1930	Loan for processing eclipse plates	NG	900	l
	Salary of computing assistant Lena Borchheim		150	m
June 1930	Loan for processing eclipse plates	NG	1,500	l
Oct. 1930	Payment for additional analysis (corona)	NG	1,000	l
Feb. 1931	Salary raise for assistant	NG	200	m
Mar. 1932	Credit for light deflection measurements	NG	1,800	l
Sept. 1932	Travel grant (Jena)	MC	100	l
1932–33	Credit for perihelion anomaly analysis	NG	10,000	l

SOURCES: Main Archive of the Academy of Sciences, Berlin; Archive of the Max Planck Society, Berlin; Federal Archive (Bundesarchiv), Potsdam Section.

NOTE: *a*, annual; *GTR*, general theory of relativity; *KWIP*, Kaiser Wilhelm Institute of Physical Research; *l*, lump-sum payment; *m*, monthly; *MC*, Ministry of Culture; *NG*, Notgemeinschaft der Deutschen Wissenschaft; *PAS*, Prussian Academy of Sciences.

[a] In 1923 hyperinflation compelled a drastic devaluation of the mark, to the rentenmark. In 1924 the reichsmark was introduced.

(see Table 5), despite the fact that Freundlich often had no more to offer that funding institution than such platitudes as the following: "I would like to thank you sincerely for the renewed appropriation of the considerable grant of 2,500 marks for our research. We have made very substantial progress in recent months, even if the fruits of some major efforts have often only been the knowledge that the problems are actually more difficult than had been previously assumed."[33]

In a 1927 booklet on the Einstein Foundation's tower telescope (*Das Turmteleskop der Einstein-Stiftung*) Freundlich gave a historical account of the naming of the Einstein Foundation, and in an effort to gloss over the failure to fulfill the tower's original purpose he underscored the very much "broader basis" on which the Einstein Institute's scientific research had been built.[34]

Other astrophysical and spectroscopic research programs conducted at the tower were much more successful than Freundlich's in proving the existence of subtle effects postulated in the general theory of relativity. The summary in Table 5 gives evidence of Freundlich's talent for raising funds not only for his own projects but also for those of his collaborators; many of these won international renown with their research and gained directorships at German or British observatories. "Under Freundlich's enthusiastic and stimulating leadership the Institute soon became a most lively centre of research and attracted a large number of visitors from all over the world. . . . The team-spirit there under Freundlich was a fine and most happy one."[35] Max von Laue's student Johannes Picht (1897–1973) and Arnold Sommerfeld's student Albrecht Unsöld (1905–1995) both worked for many years at the Einstein Institute on scholarships from the Notgemeinschaft der Deutschen Wissenschaft, and Freundlich submitted a number of successful applications to extend their contracts.[36] Unsöld's work on the theory of stellar atmospheres and other research by these students resulted in some of the most important and most frequently cited contributions of the Einstein Institute during the Freundlich era. This is confirmed indirectly in Freundlich's character evaluation of Unsöld for the Notgemeinschaft:

Dr. Unsöld is—if I may say so without exaggeration, since it agrees with Mr. Sommerfeld's view and is also justified by the reception his research has immediately found—the first genuine talent to have appeared for many years to devote himself to astrophysics; and for *both theoretical and practical German astrophysics*, which is unfortunately at an *extremely low point*, it is of direct and vital importance that this astrophysical talent be sustained.[37]

Through Unsöld's research on stellar atmospheres and line profiles, or Picht's work in theoretical physics, members of the institute became intimately associated with developments in atomic theory.[38] This again followed the example of the Mt. Wilson Observatory, where leading theoreticians like Alfred Fowler and Henry Norris Russell were invited guests and consultants. Unsöld virtually embodied this close cooperation between atomic physics and astrophysics: each year, between March and November—that is, in the astronomical observation season—he worked at the Einstein Institute, while in the winter term he lectured at Munich University, thus establishing "steady scientific contact between the Einstein Tower research and Prof. Sommerfeld's scientific circle."[39]

Aside from supporting such visiting scientists for longer periods of time, Freundlich also publicized his institute by holding lectures abroad[40] and by inviting foreign scientists to give talks in Potsdam. One example out of many is the young Indian astrophysicist Subrahmanyan Chandrasekhar (1910–95), who related his impressions of a ten-day stay in Berlin in 1931. After viewing various tourist attractions of the city, which he described as "big, clean, beautiful and magnificent," and spending a day at the Kaiser Wilhelm Institute of Physical Chemistry, he also visited the Astrophysical Observatory and the Einstein Institute in Potsdam:

He was thrilled to meet Erwin Finlay Freundlich, an astronomer and astrophysicist of considerable renown. While Chandra was familiar with Freundlich's work, he did not expect the reverse to be true. On the contrary, as Chandra noted in a letter to his father on 22 September 1931, ". . . it was such a pleasant surprise for me when I found that Prof. Freundlich not only recognized me, but was familiar with my work! He wanted me to make a small 'Vortrag' (unofficial lecture) on my work 'Stellar Absorption Coefficients.' I spoke for an hour or more on the work I had done at Göttingen. . . . Prof. Freundlich congratulated me very much on my latter work. . . . Prof. Freundlich and his wife were 'at home' to me that evening."[41]

This Fellow of Trinity College, Cambridge, future Nobel laureate, was but one of numerous contacts Freundlich had with leading astrophysicists, including, for example, Edward A. Milne, Arthur S. Eddington, George E. Hale, and Walter S. Adams. As the above quote illustrates, Freundlich was well informed about recent scientific developments outside Germany and was clearly acknowledged by prominent experts. In 1928 Freundlich was one of the few German scientists to attend the third meeting of the International Astronomical Union in Leyden. As was the case in other international scientific organizations, low attendance by

German scientists was a conspicuous remnant of the boycott in science on both sides during World War I.[42] At the commission on solar physics, of which Giorgio Abetti (Arcetri), John Evershed (Ewhurst), Hugh Frank Newall (Cambridge), Henri Deslandres and Lucien d'Azambuja (Meudon), Charles Edward St. John (Mt. Wilson), and E. F. Freundlich were members, Freundlich reported on the progress being made at Potsdam by von Klüber and Unsöld in their photometric analyses of Fraunhofer lines.[43] But he also described the problems still remaining in a comparison of his own measurements against theoretical postulations arising as a consequence of both Einstein's theory, particularly with regard to center-limb variations,[44] and Milne's theory regarding chromospheric lines. Edward Arthur Milne (1896–1950)[45] was later Freundlich's guest at Potsdam in 1932. A letter mentioning his travel plans and projecting his being "in the heart of the Einstein territory" also shows how closely Freundlich's name was associated with Albert Einstein's, even abroad.[46] In comparison, Ludendorff and his collaborators seem not to have built up international connections to the same degree—at best only with classical astronomers abroad, such as Harlow Shapley at Harvard or a few individuals connected to the project of systematically compiling all visible stars in the southern hemisphere. It should be noted, incidentally, that just as research activities at Potsdam diverged in the Astrophysical Observatory and the Einstein Tower, there was a comparable conflict in the United States over astrophysical research topics and approach. There it appeared in the form of a power struggle between the "western gang" (Mt. Wilson; i.e., Hale) and the "eastern gang" (Shapley and the Harvard Observatory).[47]

Freundlich was thus a quite capable fund-raiser and research director. Apparently he cut a smart figure on first impression, at least to the architect Richard Neutra, who described him as follows in 1921: "Einstein's assistant is a slender man whom one could describe as nearly beautiful, with an English officer's mustache, who is impatiently waiting to start his work, so he can show these earthlings—what?"[48] His personal skills left much to be desired, however. In 1921 his relationship with his mentor cooled significantly after a big row about one of Einstein's manuscripts. In Freundlich's account of the events, he had intended to sell an Einstein manuscript—with the author's consent—to help finance research at the Einstein Tower; but Einstein suddenly insisted that he had only lent the manuscript and demanded its return.[49] Arnold Berliner, editor of the journal *Naturwissenschaften*, in which both Einstein and Freundlich published from time to time, tried to mediate.[50] But he was without success. Even eight years later, after a heated board meeting during which Ludendorff exploded furiously at Freundlich, which Berliner condemned as "insulting" and "inappropriate," Einstein wrote sarcastically to Berliner in reply:

> On the one hand, I am pleased that you have such a strong moralistic streak. On the other hand, I am sorry that you waste your emotions on such unworthy objects. At any rate, I must say that though Ludendorff appears to be considerably the clumsier of the two, he nonetheless appears to be much the more decent. I for my part do not consider it beneficial for me to interfere in any way in this argument, but I will respectfully deposit your letter in what corresponds here to what you might call the files.[51]

Einstein remained distinctly aloof from Freundlich throughout the remainder of his time in Berlin. Important duties incumbent upon him, such as directing the Kaiser Wilhelm Institute of Physical Research and presiding over board meetings of the Einstein Foundation, were performed de facto by Max von Laue and others around him. Shortly before the meeting of the foundation's board of trustees in January 1929, which was to become the subject of the letters mentioned above, Einstein wrote a succinct letter to Max von Laue, officially inviting him to participate in the meeting and then continuing familiarly: "Semiofficially, I assure you that you come at your own cost, because 'sparks will fly.' I'm even looking forward to it. Logic alone isn't everything in life. You also need something for your black heart."[52]

A letter Einstein sent to Freundlich's archenemy Ludendorff—perhaps in an effort to make Ludendorff more amenable to compromise during one of the many disputes with his former assistant—illustrates how much his working relationship ·with Freundlich had deteriorated already by 1925:

> With regard to Mr. Freundlich, you know my opinion, of course. In any case I have broken off personal relations with him and could have added a few very fine "specimens" to the list of sins you reported. He counts among the very few with whom I consider such a rigid attitude necessary. But I respect his organizational achievements and act accordingly, as you have also most commendably done on the occasion of his appointment. Thus we both serve the cause, even though we value the man and the scientist little. He is not worth getting upset about.[53]

Einstein and Ludendorff were certainly not alone in finding Freundlich a difficult personality, as Figure 6 illustrates.

CHAPTER 9

The Solar Eclipse Expedition of 1929

❖❖

Despite the growing tensions between Freundlich and Einstein in the 1920's, the astronomer was able to win support for another expedition, this time to observe the solar eclipse on May 9, 1929, at Takingeun, Sumatra. This expedition, unlike those that had failed because of political circumstances (as in 1914) or weather conditions (as in 1922 and 1926), enabled the Potsdam astronomers to bring home high-quality photographic plates of the fields of stars surrounding the sun, and comparison exposures.[1] Walter Grotrian also made excellent photographs and spectrographs of the solar corona that later enabled him to solve the puzzle of the physical conditions in the outer solar regions, where temperatures are well above 1 million degrees.[2]

A precision horizontal double camera was used, which Freundlich and his mechanic, Erich Strohbusch, had especially designed for the occasion. (See Figs. 38–40.) The optics essentially consisted of two identically constructed horizontal cameras with a focal length of 8.5 m and an aperture of 20 cm, which were placed at an angle of approximately 25° to each other.[3] Both cameras received light from the same coelostat, a special construction by the Zeiss Company in which particular care was given to the precision of the automatic driving mechanism. This mechanism included an electromagnetically controlled chronometer to assure the sharpest possible stellar images.[4] Photographic plates of considerable size (45 cm × 45 cm) were attached at the focal plane of both horizontal cameras, covering a 3° × 3° field of stars scaled at 1′ = 2.5 mm. One of the two cameras was pointed at the eclipsed sun and the surrounding field of stars; the other, at a comparison field. A thermally isolated collimator with an imprinted rectangular grating was inserted in the path of the rays in front of a coelostat mirror[5] 30 cm in diameter to produce a regular grid on the

Fig. 37. Members of the Potsdam solar eclipse expedition to Sumatra, 1929: (left to right) H. von Klüber, E. Strohbusch, E. Grotrian, E. Freundlich, K. Freundlich, H. Veenhuizen, W. Grotrian. (From KB, by permission of Prof. B. Evers)

photographic plates. This grid provided an independent check for accidental changes in scale on the plates. The same procedure was repeated a few months later to photograph the same field of stars at night under identical conditions. One observer (Harald von Klüber) had to stay on location with the instruments for this purpose until the end of 1929. At this time exposures were also made on the reverse side of a glass plate. The resulting mirror-image photographs could then be placed directly on top of the other plates for comparison. In this way it was no longer necessary to take absolute star position measurements, only differential measurements of deviating star positions.[6] Out of the micrometrically determined variations in stellar position (comparison of nocturnal plate with eclipse plate), the shifts were derived by means of the following simple calculation:[7]

$$\Delta x_i = (X_{eclipse} - X_{glass})_i - (X_{nocturnal} - X_{glass})_i . \qquad (2)$$

Freundlich also used a separate parallactically mounted astrograph with a focal length of 3.4 m and an aperture of 20 cm, which was designed to produce $7.5° \times 7.5°$ photographs scaled at $1' = 1$ mm. These plates could record up to 100 bright stars each and were, moreover, well focused right to the edge of the image.[8] The astrograph was first focused on the eclipsed sun and then, without changing the photographic plate, focused on a comparison field. In all, three such doubly exposed plates could be obtained during the eclipse. The logic behind this procedure was that the stellar position measurements from the comparison field could provide corrections in scale for the photographic plate. Unfortunately, this method did not yield any usable results. The repositionings performed in rapid succession during the eclipse of only a few minutes' duration presumably caused mechanical tensions within the instrument, which led to distortions in the comparison field image that could not be reconstructed and eliminated.[9]

The complicated procedure of interpreting the results involved 100,000 individual micrometric measurements and just as many tedious calculations; it was only completed in 1930, despite the addition of Lena Borchheim to the staff for the calculation work. The photographic plates were first measured micrometrically, using a specially designed measuring device[10] at the Astrophysical Observatory in Potsdam. The raw data thus acquired were applied in the algebraic formulas given below (Eqs. 3, 4), where the terms 1–5 refer to the following potential geometric and phys-

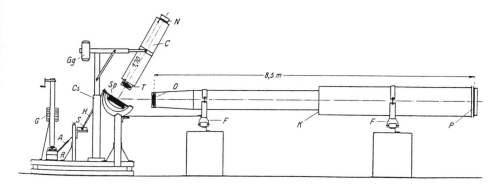

Fig. 38. The Potsdam horizontal double camera with coelostat and driving mechanism, first used successfully during the 1929 solar eclipse expedition, schematic side view: G, regulator drive weights; R, regulator with electrical second control; A, connecting arm, regulator screw; S, housing for driving screw; H, lever from screw to mirror; Sp, flat mirror, 400 mm; Cs, collimator column, rotatable along vertical and horizontal axes; Gg, counterweight; C, collimator tube with grid (N) at the focal point of the condenser lens (T); K, camera tube; O, lens; P, cassette with photographic plate; F, guideways. (From Freundlich 1930a [317], by permission of Springer-Verlag; see also Freundlich et al. 1931b [7])

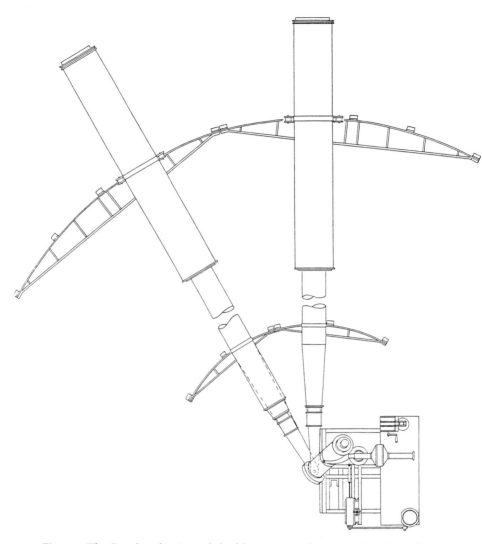

Fig. 39. The Potsdam horizontal double camera, schematic view from above, scale approximately 1:40. (From Freundlich et al. 1931a [177], by permission of Springer-Verlag)

ical influences on the stellar shifts: the influence of the plate's position in (1) the measuring instrument and in (2) the camera were computed, in relation not only to (2a) the different tilts but also to (2b) a change in focal length, incidental astronomical influences such as (4) aberration and (5) refraction, and not least the possible physical influence of (3) light deflection within the gravitational field.[11] Freundlich, von Klüber, and Albert von Brunn (1880–1940) interpreted the raw data obtained from Equation 2 (above) using these formulas for the x or y components of the stellar coordinates (difference: daytime minus night photo):

$$\Delta x = \underbrace{a + by}_{1} + \underbrace{px^2 + qxy}_{2a} + \underbrace{Sx}_{2b} + \underbrace{Ex/r^2}_{3} + \underbrace{Ax}_{4} + \underbrace{Rx}_{5} \quad (3)$$

$$\Delta y = \underbrace{a - bx}_{1} + \underbrace{pxy + qy^2}_{2a} + \underbrace{Sy}_{2b} + \underbrace{Ey/r^2}_{3} + \underbrace{Ay}_{4} + \underbrace{Ry}_{5}. \quad (4)$$

a, b, p, q, S, and E are the unknowns, all but S determined through least-squares calculations. A and R are calculated individually from the data attached to each of the photos at the time of exposure, such as temperature,

Fig. 40. Freundlich (center) in front of the horizontal double camera used to record star positions to detect light deflection in the sun's gravitational field during the solar eclipse in Sumatra, 1929. (From Freundlich et al. 1931a [176], by permission of Springer-Verlag)

Fig. 41. Solar eclipse photo (one of the four taken during the Potsdam expedition in 1929) including corona and surrounding stars (numbered). (From Freundlich et al. 1931a [183], by permission of Springer-Verlag)

barometer reading, exposure time, and star coordinates. In this optimization, involving 36 equations for the 18 stars under examination, separating the two unknowns S and E posed a particular difficulty. This was because

as the central part of the plate is taken up by the moon or by the eclipsed sun and its surrounding corona, stars only appear at distance r, which is significantly larger than 1; and thus the coefficients of E become correspondingly smaller than 1. As a result, the weight of determining E is small, while that of S, whose coefficient increases sharply in relation to the constantly rising number of stars as you move away from the rim, becomes correspondingly large. . . . So from the available systematic shifts of eclipse versus night [photo] stellar [coordinates], the heavily weighted S has the tendency to become inflated at the expense of the main unknown E, of small weight, so that the latter runs the risk of ending up being too small.[12]

In later papers Freundlich defined this mathematical coupling of the two unknowns S and E by the relation[13]

$$\Delta E \simeq r^2 \, \Delta S. \tag{5}$$

Thus, for example, with an average stellar distance of $5r$, any error in determining S affects the resulting light deflection value by a factor of 25. The Potsdam observers consequently assigned great weight to the scale value S based on this relation, because it could be derived *independently* from an interpretation of stellar positions using the comparison grid imprinted on the photographic plates. Such a comparison directly after the eclipse revealed that "in the images on the two plates . . . only very rarely does the difference between the actual length and the proper length, composed of segments of ten grating ruling intervals, exceed the total of $\pm 5\mu \simeq 0''.1$."[14] This grid comparison demonstrated primarily that other conceivable systematic sources of error could be eliminated. These included contraction of the exposed photographic plate while the emulsion was drying, caused by the extremely uneven exposure of different areas.[15] Figure 42 illustrates that the distances between the grating lines were nonetheless also subject to negligible yet irregular fluctuations.

The fixed star positions during the solar eclipse were first compared against their normal positions on von Klüber's night sky photograph without taking into account the "Einstein Effect" $\sim E$, in order to obtain the difference in scale values for S:

$$\Delta x = a + by + px^2 + qxy + Sx \tag{6}$$

$$\Delta y = a - bx + pxy + qy^2 + Sy. \tag{7}$$

In a second series of calculations the influence of aberration and refraction from the earth's atmosphere, which had to be computed separately for each photograph and for each stellar coordinate, was removed from the resulting shifts Δx and Δy.[16] Then a final calculational optimization of the observed field of stars was carried out through incorporating the term Ex/r^2 in Equation 3 (above). Figures 43 and 44 illustrate the results of this laborious, two-step data evaluation procedure. A comparison reveals that, disregarding relativistic light deflection in the solar gravitational field, the stellar position shifts produce a pronounced scattering in the distribution of displacement vectors.[17] Incorporating the relativistic correction in the form of the terms $E/r_i \cdot x_i/r_i$, an approximately centrally symmetric distribution of the displacement vector orientations results. Moreover, their

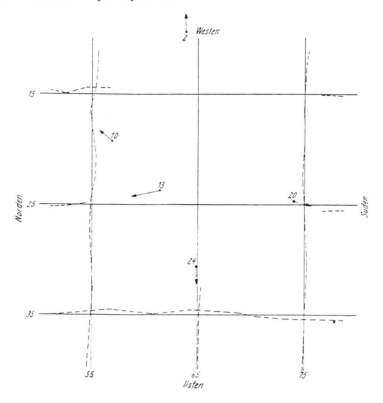

Fig. 42. Superimposed gratings of two photographic plates from the Potsdam solar eclipse expedition of 1929, showing irregularities in the imprinted comparison grids. For comparison the displacement vectors resulting from light deflection for a few stars are recorded in the same scale. Scale of the original image: 1 mm = 0″.17. (From Freundlich et al. 1931a [182], by permission of Springer-Verlag)

magnitude falls as distance r increases away from the sun in accordance with the $1/r$ law.

The deflection angle in the immediate vicinity of the sun's rim could not be measured because of the corona's brightness, but it could be easily determined from the data's dependence on the $1/r$ relation. (See Fig. 44.) An analysis of the four plates by Freundlich and his collaborators yielded the values 2″.25, 2″.17, 2″.61, and 1″.81. With extrapolation from the final measurements of the 18 relevant stars in accordance with the hyperbolic law required in relativity theory, that averaged out to 2″.24 ± 0″.10 (as

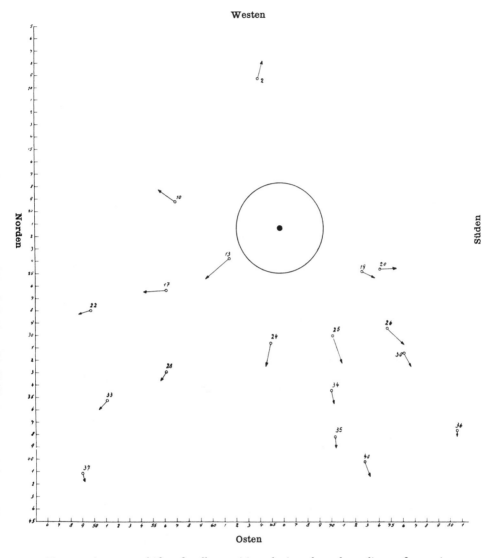

Fig. 43. Apparent shifts of stellar position during the solar eclipse of 1929 in relation to their normal positions, including relativistic light deflection. (These are the mirror image of Fig. 41.) All vectors have been exaggerated by a factor of 1,000. (From Freundlich et al. 1931a [184–85], by permission of Springer-Verlag)

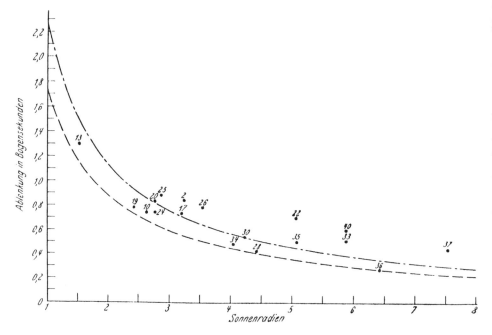

Fig. 44. Apparent shifts of 18 star positions recorded during the solar eclipse of 1929, plotted in relation to their distance from the sun's center (mean values from the four plates of the eclipse field). (From Freundlich et al. 1931a [191], by permission of Springer-Verlag; see also Freundlich et al. 1931b [35] for a separate analysis of each photographic plate)

opposed to $1''.75$ according to relativity theory).[18] It should be noted here that the closest suitable star was approximately 1.5 solar radii away from the sun's center; the farthest, 7.5 solar radii away. In the intermediate area of about 5 solar radii, deflection measurements already deviated substantially from the predictions of relativity theory, with a tendency toward greater values than expected. Einstein was quite irritated by the large discrepancies in the values for E (between $1''.6$ and $2''.6$, with an average error for each individually determined value of $\pm 0''.2$). Freundlich attempted to placate him with the following explanation:

The apparently drastic variation in values for E . . . is actually not excessively great. Normally we must always reckon with deviations from the mean value of up to twice the average error amount; thus in our case this means up to $0''.40$. . . . There is no obvious dependence of the value for E upon the length of exposure— 40 sec, 90, 60, 40 sec—and a dependence upon the sequence of photographs is equally unlikely. It would have been conceivable that the coelostat mirror deformed progressively as the eclipse phase proceeded to produce the same errors. This also is not apparent.[19]

He also dismissed Einstein's query about the possible influence of anomalous refraction from the earth's atmosphere, referring to Robert J. Trümpler's estimate of this effect as minimal ($< 0''.001$).[20] But Einstein was not convinced.[21]

Freundlich, von Klüber, and von Brunn claimed a remarkably small margin of error, only $0.''10$, for their unexpectedly large result—disregarding the considerable discrepancy with the value postulated in relativity theory for light deflection at the sun's rim. This was far lower than all previous observations and also well below the variation in the values from the four individual plates ($0''.40$) from which they had obtained their results through averaging. Their interpretation culminated in the statement:

Thus it is hardly possible to doubt that our series of measurements does not agree with the value $1''.75$ claimed by the theory. The question of whether our amount might be invalidated by a systematic error cannot be determined finally as long as the possibility remains that unknown influences affect the results. . . . In the meantime the unavoidable conclusion is therefore that the deflection of light in the proximity of the sun is found to be larger than the value predicted by relativity theory.[22]

The comparison field data, which were evaluated using exactly the same method as for the reduction of the eclipse photographs, corroborated these results. This check gave no indication of systematic effects—specifically, no radially symmetric shifts, and no anomalous refraction phenomena in the comparison field. The resulting shift vectors were very much smaller than those in the eclipse field.[23] Since Freundlich's findings contradicted previous solar eclipse expedition results, Freundlich and his collaborators ended their analysis with a discussion of these earlier findings. Their essential point both on Eddington's and Crommelin's 1919 measurements and on Campbell's and Trümpler's 1922 data[24] was not an easy one to accept. It again amounted to calling into question the *communis opinio* that had grown out of the extensive publicity in the preceding dozen years about the British expedition of 1919, which was celebrated as a clear confirmation of the general theory of relativity:

In reality the first determination of light deflection by the English expedition in 1919 already provided an unambiguously larger value for E than relativity theory had predicted. The data observed by the English expedition of unquestionably the greatest weight, namely the data obtained with the 4-inch lens of a 6 m focal length, yielded $E = 1''.98$ from seven plates, each with seven stars close to the sun. But this divergence was not deemed realistic when some years later the Lick expedition measurements arrived at the exact theoretically predicted value. It can be shown, however, that an error in the reduction procedure simulated this total confirmation of the theory of relativity, and that through a correct reduction of the value, $E = 2.''2$ results. Furthermore, it can be shown that the first English observations also support a light deflection value greater than the value then calculated of $E = 2''.00$, closer to $2''.2$.[25]

Elsewhere Freundlich and his collaborators strengthened their argument further by referring to the quasi-linear coupling of the scale value S and the Einstein value E in the least-squares calculations. (See above, text with Eqs. 3 and 4.) They were the first to work this out clearly, because the horizontal double camera enabled them to include an independent determination of the scale value in their measurements: "That is why, with the exception of the observations made with our horizontal camera, up to now all measurements of light deflection lack the methodological basis guaranteeing the quantitative reliability of their determinations."[26] They also presented this provocative revision of earlier measurements graphically, plotting the newly computed shifts in relation to the distance from the sun on the basis of the previously published raw data from the English and American expeditions and their own. (See Fig. 45.)

Freundlich's interpretation of the data naturally did not remain uncontested for long: Ludendorff and, independently, Trümpler also pointed to the large source of error caused by the asymmetrical distribution of bright stars near the edge of the sun. Using another reduction method Trümpler then arrived at the value $1''.75 \pm 0''.13$, and Ludendorff at the mean value $1''.95$. Danjon's calculations came to $2''.06$; Mikhailov's, to $1''.96 \pm 0''.11$. Jackson, suspecting an error in the determination of the scale value S, proposed $1''.98 \pm 0''.20$. Thus Freundlich's own results were not immune to reanalysis.[27]

Besides, in considering Freundlich's daring theses it was presumably of some significance to his colleagues that he had already become embroiled in a bitter controversy in 1915 and 1916. In the Lick Observatory archives there is a translation by William Wallace Campbell (1862–1938) dated October 1931 of Hugo von Seeliger's anti-Einstein article of the same month. Campbell had surely prepared it only to be able to share von Seeliger's vehement denunciations of Freundlich with his English-speaking colleagues. A letter by Campbell to Trümpler is in the same vein. He

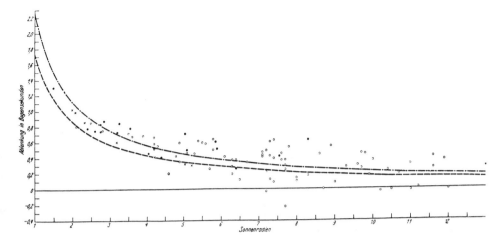

Fig. 45. Comparison of light deflection results of the English (1919: ✕), American (1922: ◯), and German (1929: ●) solar eclipse expeditions. The dashed curve represents the hyperbolic decrease as predicted by the theory of relativity; the dotted dashed curve, the decrease found in the Potsdam expedition measurements assuming a hyperbolic decrease. (From Freundlich et al. 1931a [197], by permission of Springer-Verlag)

reports on Freundlich's attacks on their own researches of 1923–24 and 1928 and then continues:

Confidentially I am also telling Dr. Prichett [president emeritus of the Carnegie Institution] that Dr. Freundlich was compelled to resign from the Berlin-Babelsberg Observatory about the year 1915 by its Director, Hermann Struve, because of Freundlich's trickery and dishonesty in scientific matters, as shown up by Seeliger in the [Astronomische] Nachrichten vol. 202, page 83. To give you the right angle on Freundlich, I advise you to read that article, if you have not already done so. The whole incident was described to Dr. Moore and others here a fortnight ago by Professor Georg Struve, an astronomer in Berlin university and Babelsberg observatory, who has been making observations with the Lick telescope during the past three months. He also remarked that Freundlich has no standing whatever with German astronomers.[28]

Despite Einstein's strained relations with Freundlich and the latter's rapidly decaying reputation in the scientific community, in 1931 Einstein did consider a possible correlation between the anomalies Freundlich claimed and his own attempts at a unified theory of gravitation and electricity with the connected modified field equations.[29] But this moment of doubt was only brief:

In the last few days I have found impressive indications that a general theory of relativity based upon a parallel structure cannot be reconciled with the facts. Thus, from a theoretical standpoint, an exact validity of the old equations becomes very likely.[30]

And soon thereafter:

The old gravitation equations are based not just upon the relativity postulate, which in itself a discerning person could hardly doubt, but also upon the assumption of a relatively far-reaching distinction between the law of gravitation and the law governing the total field. This latter assumption has no basis. It is precisely for this reason that it is of such great importance to consider accurately the significance of the experimental results.[31]

Unfortunately, Einstein's mathematical speculations never reached the point where a testable prediction, such as a modified light deflection value, could be advanced.[32] In 1932 Einstein seems to have adopted the majority view and described Freundlich's findings as the result of an "erroneous calculation of the experimental results. . . . Through correct calculation good conformity with the theory is again reached."[33]

The debate between Ludendorff and Freundlich in 1932 in the *Astronomische Nachrichten* illustrates how bitterly the opponents fought over the interpretation of the Potsdam data.[34] Eventually Freundlich and von Brunn conceded that considerable improvements in the mechanical and optical precision of the instruments would have to be made (e.g., the relative changes in focal length $< 10^{-6}$) and that future plates would have to obtain the then quite utopian scale accuracy of $\Delta S < 10^{-5}$ to produce suitable data.[35] In 1937 von Brunn and von Klüber went further by demonstrating that, even with such optimized equipment as Freundlich had himself used in Sumatra in 1929, because of obscuring by the corona and other unavoidable systematic optical sources of error the accuracy of the results "cannot be improved substantially over that reached at the time of the eclipse."[36] But despite the fact that all this research was at the extreme limit of technological precision for that time, Freundlich continued to be convinced that his results were generally viable, despite being at odds with Einstein's light deflection prediction. The tendency of his findings seemed to be supported not only by his very careful determination of the scale values through comparison against an independent ruled scale but also by later observations, which also tended to yield large values for light deflection at the sun's rim.[37]

CHAPTER 10

Clashes Between Freundlich
and Ludendorff

❖

Administrative bickering compounded the professional tensions at Potsdam. The exact legal status of the Einstein Institute in relation to the Astrophysical Observatory, which had been devised by the Einstein Foundation's legal adviser, Ludwig Ruge, was at best ambiguous. In order to decide whether the Einstein Institute was eligible to send its own representative to a meeting of the International Astronomical Union, in 1932 the German astronomers' committee inquired at the Ministry of Culture whether Freundlich's institute was an autonomous institution. The committee chairman received a Delphic reply: the Einstein Institute in Potsdam was "neither independent nor dependent" and the institutes in Potsdam promoting astrophysics carried the common title "Astrophysical Observatory and Einstein Institute."[1] Officially the "scientific organization created by the Einstein Foundation" was not an independent research facility, but "in the administration of the supervisory board, it [was] a part of the Astrophysical Observatory." In any case all the ministry's decrees, even those affecting the Einstein Institute alone, and all the institute's working and annual reports and funding applications had to land on Ludendorff's desk at the Astrophysical Observatory in Potsdam. Conversely, applications by the Astrophysical Observatory that involved the Einstein Institute in any way had first to be approved by the Einstein Foundation's board of trustees, which was responsible for supervising the disbursement of all funds collected. (This board is not to be confused with the Astrophysical Observatory's board, even though many of the scientists were members of both boards: see lists above in Chap. 6 and in text below.) Einstein was lifetime chairman of the Einstein Foundation's board of trustees. The other board members included the lawyer Ludwig Ruge, a representative of the German industrial association Reichsverband der Deutschen Industrie (until July 1925 its managing director, Rudolf Schneider; then Max Fischer of the Zeiss Company), and two representatives from the Ministry of Culture (Assistant Secretary Hugo Andres Krüss and Wilhelm Westphal).[2] Freundlich had signed his directorship contract with this board in 1922, and he considered himself accountable to this body alone.

Starting in fiscal year 1925 the Einstein Institute had a separate budget that drew on the Astrophysical Observatory's available operational expenses fund, to which only the Einstein Institute trustees had access. Nonetheless, approval still had to be sought from the director of the Astrophysical Observatory for all disbursements that did not qualify as operating costs (e.g., the purchase of instruments, which constituted an increase in real assets) "to avoid duplication of acquisitions." Between 1920 and 1931 the members of the Einstein Foundation's board of trustees were:[3]

From December 1920:	Albert Einstein, Erwin F. Freundlich, Hans Ludendorff, Gustav Müller
From June 22, 1922:	Einstein, Freundlich, Ludendorff, Müller (d. 1925), James Franck, Carl Bosch, Rudolf Schneider, Jeidels, Ludwig Ruge
Additionally from November 5, 1924:	Friedrich Paschen (president of the Physikalisch-Technische Reichsanstalt)
Around 1931:	Einstein, Max von Laue, Erwin Schrödinger, Emmanuel von der Pahlen
January 1, 1932:	Transfer of the Einstein Institute to the Prussian state.

The entangled administration of the two institutions provided a constant source of friction between Freundlich, head of the Einstein Institute, and Ludendorff, the Potsdam Astrophysical Observatory's director from 1921 to 1939.

(Friedrich Wilhelm) Hans Ludendorff (1873–1941)[4] came from a family "in which science and soldiery were part of its tradition." His great-grandfather, General G. F. Tempelhof, "had also, aside from his military activities, devoted himself to astronomical questions."[5] His younger brother, General Erich Ludendorff, earned his reputation as a commander in World War I; after Germany's humiliating defeat, he helped create the myth of the "stab in the back" (*Dolchstoßlegende*) and later entered politics, cooperating with Hitler. Hans Ludendorff submitted his doctoral thesis at Berlin University in 1897 on a thoroughly traditional topic in classical position astronomy.[6] On being engaged at the Potsdam Astrophysical Observatory in 1898, the young man slowly worked his way up from a simple assistantship to become its director. His was the systematic approach. For example, until 1899 he compiled a photographic sky map, which required painstaking measurements and reductions of the plate material on into late 1900. With this background he would naturally care little for the direction Freundlich's researches had been taking since 1913 in investigating the theory of relativity experimentally. He resented the external support Freundlich was enjoying and felt passed over. Freundlich's high-handedness in usually not even consulting him on administrative matters was not conducive to good relations; neither were their political differences. Freundlich once wanted to introduce him to Max Delbrück (1906–81), a new staff member who later became famous for his pioneering work in molecular biology:

When Hans Ludendorff heard the name Delbrück, he immediately snatched his hand away and asked whether Max was the son of the historian. When Max said that he was, Ludendorff turned on his heel and disappeared into his office, slamming the door behind him. Later he accused Freundlich of intentionally insulting him by hiring the son of a man who had intolerably insulted his brother. . . . It was many weeks before [Ludendorff] calmed down.[7]

In some cutting book reviews the elder Delbrück had accused General Ludendorff in 1920 of "shallow theoretical thinking," "incredible ignorance," and even mismanagement of the war: "We were incapable of preventing the World War, . . . but we . . . could have ended the war otherwise, had Ludendorff been a different man."[8]

Ludendorff put every obstacle he could in Freundlich's way to hamper his projects or even to subvert them, and he took every opportunity to blacken his rival's reputation.[9] In light of the abiding enmity between the two, passages like the following, from an obituary of Ludendorff, take on a quite different color:

Fig. 46. Hans Luden-dorff (late photo). (From Kopff 1941: 225)

As head of the observatory Ludendorff became well liked for his genuine and honest humanity and for his high regard for the value of scientific research. In keeping with this attitude, he gave his coworkers free rein to develop their scientific talents fully through research of their own choosing. He spared no pains in protecting the observatory's welfare and scientific standing, and he had to fight many a battle on its behalf.[10]

From Ludendorff's point of view Freundlich was certainly not conducting constructive research and thus could not be granted the liberties other members of the staff apparently enjoyed. Moreover, he could only perceive the young relativist astronomer's activities as undermining the reputation of the observatory.

Freundlich's less than diplomatic dealings with others (see Fig. 6) only fanned the flames in daily clashes between two men of such different principles. He occasionally tried consciously to circumvent Ludendorff by avoiding official channels.[11] Examples abound in the files, and we will have to limit ourselves to a few highlights.

In January 1925 Ludendorff complained directly to some administrative officials about unduly high consumption of electricity during the preceding Christmas holidays. Freundlich countered: "I do not consider it appropriate that such uncontrollable intrigues by some official or other come to my attention only weeks later, when any kind of clarification of the facts has become impossible, without my being given the opportunity to express my opinion beforehand."[12] In 1931 Freundlich planned to test the reflectivity of mirrors produced with a new silver-plating technique. He requested the financial support of the Ministry of Culture in purchasing at about half its market value (about 14,000 reichsmarks) a used reflecting telescope that the photographic engineering laboratory at the Charlottenburg Polytechnic in Berlin was offering for tests of the new method. When the Einstein Institute's application came into Ludendorff's hands, he commented in an enclosure to the minister:

> The granting of so significant a sum for the purpose indicated is in my opinion not advisable. The tests planned by Prof. Freundlich . . . can be performed at much lower cost in the laboratory. . . . It must also be noted that the time that the Zeiss Company allowed for payment for the tower telescope's 60-cm lens will soon be up. The state will then have to part with very considerable sums. If the advance of further funds is at all conceivable under present financial conditions, then I would like to request that they be applied toward improving the Astrophysical Observatory's outdated instrumentation. I last pointed out the urgent necessity of such improvements in my proposals for the 1931 state budget. Specifically, with its tower telescope the Einstein Institute section of the observatory possesses a modern instrument, but as yet it has produced only very few results. Right now the older sections of the observatory should primarily benefit from instrumentation improvements.[13]

This "expert opinion" shows that clearly at the end of the day both parties were fighting over the appropriation of funding, which had become extremely scarce with the global economic crisis of 1929. In their efforts to expand their own institutions under these circumstances, Freundlich and Ludendorff were necessarily pitted against each other. Ludendorff did in fact have a point: since the establishment of the Einstein Foundation in 1920, very much more money had been spent on instruments for the Einstein Institute. In contrast the Astrophysical Observatory's instruments had remained unchanged, and thus threatened to become obsolete. Even his swipe at the institute's meager results despite its modern equipment was, as mentioned above, certainly not completely unjustified in view of the tower's original purpose. On the other hand, at that time the Astrophysical Observatory itself was conducting only conventional research, so that there was no urgent need for new apparatus.[14] However justified either argument was, the above excerpts from the files fully reflect a cutthroat rivalry over financing. It was a small step from unrestrained deprecation of the opponent's projects to outright defamation.

In July 1928 there was a quarrel over a new budget, which Freundlich had carefully timed to be brought before the Einstein Foundation's board of trustees shortly before Ludendorff's departure on a trip abroad. Though denied the opportunity to give his opinion on it officially, Ludendorff still managed to put on record "certain reservations against the content and wording of the budget" and otherwise to protest against the timing of its submission. He expressed his indignation to Einstein, describing this course of action "as extremely inconsiderate—to put it mildly," and also questioned Einstein's own actions after receiving a letter from him stating, as Ludendorff described it: "in unmistakable terms your opinion of Freundlich as a person and a scholar." A few years later Ludendorff was unprincipled enough to send a copy of that letter to a senior executive officer at the Ministry of Culture, "with the request kindly to take note of its contents."[15] Whatever the reasons why Einstein made derogatory comments about Freundlich, he had made them in a strictly private letter to Ludendorff; yet it was used to discredit Freundlich at the highest levels. The relationship between the two institute heads could no longer be simply described as a professional difference of opinion: they had become bitter enemies.

Freundlich's and Ludendorff's diverging spheres of activity and preferred working methods only reflect the conflict of their personalities. Freundlich devoted his time in the environs of Berlin almost exclusively to looking for an experimental verification of the general theory of relativity, which he was firmly convinced of at least until 1931. Ludendorff, however, was equally certain of the invalidity of the theory and consequently only reluctantly conceded any time for testing it. Ludendorff took advantage of the new interest in solar eclipses fanned by the sensational press reports in 1919 to win financial support for his analysis of the shape of the sun's corona, which, like Einstein's postulated light deflection in a gravitational field, could be measured satisfactorily only during a relatively rare total eclipse. He undertook his own expedition to Mexico for the total eclipse on September 10, 1923, excluding any members of Freundlich's staff. He was accompanied by A. Kohlschütter and the senior mechanic W. Herrmann, both from the Astrophysical Observatory in Potsdam, together with Prof. R. Schorr and F. Dolberg from the Hamburg Observatory.[16] Arnold Kohlschütter (1883–1969), a student of Karl Schwarzschild's, had done research at Mt. Wilson from 1911 to 1914 and was hired as an observer at the Astrophysical Observatory in 1918. On Ludendorff's expedition he was responsible for repeating the light deflection measurements.[17] Ludendorff was granted for this expedition alone

Fig. 47. Einstein in
conversation with
Ludendorff. (By
permission of the
Bettmann Archive,
New York)

2 million marks (during the inflationary peak) by the Kaiser Wilhelm
Institute of Physical Research.[18]

Unlike most of Freundlich's expeditions, Ludendorff's observed the
eclipse under favorable conditions: "The sky all around the sun was completely clear and a deep blue during the total eclipse."[19] Although Ludendorff's resulting calculations of the intensity curve of the continuous corona spectrum were published, Kohlschütter's results apparently never
were. The fact that Kohlschütter was appointed professor of astronomy at
Bonn University in 1925, becoming simultaneously director of the observatory there, may have been the reason for this. In 1926 or 1927, furthermore, he was appointed head of the German astronomical observation
station for the southern hemisphere at La Paz, Bolivia, at which post he
remained until 1946; perhaps he never found the time to interpret his
expedition data.

Ludendorff's research in Potsdam also focused on the traditional questions raised since the time of Hermann Carl Vogel, before the turn of the
century, about stellar radial velocities, spectroscopic binaries, and variable

Fig. 48. Ludendorff (foreground, at right) and colleagues during the expedition to Mexico for the solar eclipse on September 10, 1923. (From *Berliner Illustrierte Zeitung*, April 27, 1924, by permission of Ullstein Bilderdienst)

stars. From 1930 on he also wrote historical studies in astronomy[20] and compiled comprehensive star catalogs, particularly of the southern hemisphere, for which purpose he established in 1926 the German observation station in Bolivia that he built up together with Kohlschütter.[21] Such cataloguing, the most classical task in astronomy since its beginnings, was Ludendorff's most characteristic research activity. This is confirmed in a remarkable document that the psychologist Paul Plaut reprinted in an appendix to his *Psychologie der produktiven Persönlichkeit* (Psychology of the creative personality). Plaut had distributed a questionnaire to a number of scholars on their research methods and their motivations. Ludendorff's response was:

My work focuses on the collection and organization of (observational) results insofar as these make up the preliminary stages to a discussion of the phenomena. This discussion is what is *essential* to me, and it is what gives me the most pleasure. A great amount of the latter (which in my opinion plays a particularly important role in research work) comes in my case when I succeed in deducing new facts out of the collected or ordered material or when I succeed in correlating phenomena that had to that point been considered unrelated or when I can explain these relations. In this sort of work naturally a lot of *singular* results are obtained, which also bring satisfaction.[22]

Ludendorff thus found satisfaction in collecting, ordering, and evaluating new data, as opposed to reevaluating existing data in a new way, as Freundlich repeatedly undertook to do. As a coworker phrased it in an obituary: "Ludendorff was an astronomer at heart and in soul. His successes were based on his great working capacity, the solidity of his knowledge, and his masterfully critical evaluation of existing observational data. He was less disposed to speculative considerations."[23] Ludendorff's self-critical response to Plaut's question about the importance of methodological reflection in his work was: "I have little interest in method and consider it a shortcoming of my personality as a scientist. When a new methodological element becomes necessary in solving a scientific problem before me, I regard the development of the method merely as an unavoidable necessity. I find the essence always to be the positive results of the work."[24] The contrast between this classical astronomer and the "great Einstein enthusiast" Freundlich could hardly have been greater.[25] The former sought building blocks to add to the structure of knowledge that he considered basically complete, while the latter tested the stability of its very foundations. Here is Ludendorff himself on the cumulative model of the advancement of knowledge: "My research grows quite absolutely out of the wish to fill in gaps or to continue to build on science. The desire to correct false views has always played only a very secondary role for me."[26] Even though Freundlich's response to Plaut's questionnaire was unfortunately not published,[27] it is clear that he did not adhere to such a scientific ideal, that he aimed at more than just fixing a few holes, because the system of doctrines on which astronomy and astrophysics was based would have to be rearranged if Einstein's hypotheses were to be proved true.

Tensions at the tower heightened in 1931. According to the Einstein Institute's founding articles it was to be transferred into the hands of the government on October 1, 1931, ten years after its formal establishment.

Freundlich's institute thus stood to lose its independence permanently and to be incorporated fully into the Astrophysical Observatory, under Ludendorff's direction. About six months before the fateful date the board members of the Astrophysical Observatory received a suggestion from the Ministry of Culture:

One of the donors has expressed the wish that in acknowledgment of the history of its origin it retain its independence also as a state institution and that it not be incorporated into the Astrophysical Observatory, as would suggest itself. Out of regard for the tower's continued development I am considering fulfilling this wish and guaranteeing in the future as well the de facto independence that the Einstein Tower has been enjoying up to now.

Of the persons working at the Einstein Tower only Prof. Freundlich, as director of the tower, would have to leave the Astrophysical Observatory's staff. All other persons working at the tower would remain attached to the Astrophysical Observatory by virtue of their positions.[28]

At this time the members of the Astrophysical Observatory supervisory board included Einstein, Max von Laue, Erwin Schrödinger, Walther Nernst, and Friedrich Paschen, president of the Physikalisch-Technische Reichsanstalt. They had all been reappointed on April 1, 1931,[29] and all aligned themselves behind Einstein. This left Ludendorff with absolutely no chance to take over the reins of the Einstein Tower, despite the otherwise auspicious moment. Nevertheless he raised doubts about "future staffing changes" in a preliminary compromise proposal that guaranteed the Einstein Institute one department head and two other permanent positions that would be accountable to the ministry. Ludendorff's counterproposal guaranteeing in general terms "no change for the worse" was rejected by Freundlich, however,[30] who justifiably found it too loosely worded and lacking adequate security for his coworkers—not to mention for himself. At the end of April Max von Laue, who had been directing the negotiations, proposed to the Ministry of Culture a reorganization of the Einstein Institute

in a form that guarantees the Einstein Tower one department head position and two permanent positions. The gentlemen [undersecretaries Richter and Leist] have approved the suggestions and are pleased that it appears possible to keep the peace in Potsdam in this way. They had reservations only regarding the position in which the two individuals filling the permanent positions at the Einstein Tower were to be subordinated within their discipline to the director of the Astrophysical Observatory.[31]

When the ministry eventually came up with a solution that seemed acceptable to both the board and Ludendorff, von Laue urged Freundlich not to continue to press for "an even stronger emphasis on [his] independence as head of the Einstein Institute in the new statute," because it would only jeopardize the entire settlement reached:

Please consider: it is up to the ministry how this new statute will be interpreted insofar as it concerns the relationship between the ministry, you, and Ludendorff. Mr. Leist will contact you soon and assure you that the ministry shares your interpretation of the statute entirely. So please, do not raise any objections! Otherwise you endanger everything.[32]

It is remarkable how much support Freundlich found for his demands to guarantee his own independence and that of his staff among the Astrophysical Observatory board members, at the Reich Interior Ministry, and at the Notgemeinschaft der Deutschen Wissenschaft, despite his now strained relations with Einstein. In April he complained to the Viennese physicist Samuel Oppenheim (1857–1928), who was on close terms with Einstein:

In a few months the institute that I created in his honor is to be taken over by the ministry. *One word from him* would suffice, saying that he would welcome it if the institute were able to retain its previous scientific independence and if I would be assured of what I had created in ten years of the devil knows how arduous work. But Einstein does not seem to be at all aware of the possibility of lending a helping hand here.[33]

Confident that his message would be passed straight on to Einstein through someone with whom Freundlich had not yet fallen out, he continued: "If I do not come through on my own in this matter, then I will take the opportunity available to me of going abroad. . . . With the frosty lack of appreciation here it can only get warmer up north."[34] Freundlich's strategy worked. As a self-described "honest friend of all three participants: that is, you [Einstein] and Freundlich and science as well," Oppenheim sent Freundlich's letter immediately on to Einstein, requesting that he read with an open mind the passages he had underlined and that he help Freundlich out.[35] Einstein must have done so, because in mid-April Freundlich thanked him personally for having "taken such great pains in safeguarding my institute," adding his express hopes that the projected settlement "will satisfy both sides and that my department's scientific development will also continue to meet your standards."[36] But he could not resist mentioning his foreign job opportunity to Einstein as well. He had been offered a chair in astronomy and the directorship of an observatory—at Oxford, no less. He used this trump card well, playing it off against the prospect of the Einstein Institute's certain decline as a result of his departure:

That is why a satisfactory settlement of the local conditions is so important to me; it would certainly not be easy for me to leave what I have built up here, which in all probability "would go to rack and ruin," as Kopff put it, if I were to leave. I have not said no to Oxford, though, and am waiting to see how the negotiations continue to develop.[37]

Freundlich's threat to accept a professorship abroad has another dimension as well. Unlike his more successful brother,[38] for example, and some of his own coworkers at the Einstein Tower (e.g., Walter Grotrian), Erwin Freundlich was never listed in the course catalog of either of the universities in Berlin during his years there. His attempts to obtain the necessary qualification for an academic career by submitting a postdoctoral (*Habilitation*) thesis, the first as early as 1920, foundered against the opposition of Ludendorff, among others. However, Freundlich acquired the title of professor upon his appointment as head observer of the Einstein Institute on January 1, 1922.[39] Thus formally he gained the maximum that could be expected of so troubled a career.

CHAPTER 11

Political Transition and Exile

The National Socialists' seizure of power in Germany in 1933 was a decisive turning point in other spheres than the political. It led to the forced exile of Freundlich, Mendelsohn, and Einstein, and it brought about a reorientation of research at the Einstein Tower, which soon lost not only its name but also its original function. We will naturally have to forgo a complete description of the political events that took place in this year in Germany. The events leading up to the burning of the Reichstag in Berlin, the last free parliamentary elections, the rapid exclusion of opposition parties, and the immediate prohibition of the German Communist Party based on the Empowerment Law of March 24 and the subsequent specific ban of July 14 are all too familiar. The political purge was extended to government employees, primarily through the Law for the Restoration of the Professional Civil Service of April 7, along with its implementation regulations. Accordingly, "officials not of Aryan origin" or those who could not "guarantee through their present or past political activities that they support the national state at all times without reservation" were to be retired. Frontline veterans of World War I and their dependents were at first exempted, but this provision was later retracted. With the Law Against the Overcrowding of German Schools and Universities of April 25 the way was prepared for political meddling in the previously autonomous area of education. This was accelerated by strict implementation of the authoritarian *Führerprinzip* and the exercise of political control in all aspects of public and academic life.[1] On May 10 prohibited books were burned throughout Germany under the radical action "against the un-German spirit": Freud, Tucholski, and Heinrich and Thomas Mann were forced to emigrate along with many other eminent intellectuals who had lost favor under the National Socialists.

Erich Mendelsohn, the "oriental from East Prussia," as he ironically called himself, was farsighted enough to forecast that "the destruction of the Reichstag was only the beginning" and that "when the time was ripe, the stirrup holder Papen and Hugenberg would be eliminated as well."[2] He emigrated at the end of March 1933 via Amsterdam to England. In 1935 the enterprising architect opened another office in Jerusalem; he moved to Palestine in 1939, where he designed some major architectural works, including buildings for the University of Jerusalem on Mount Scopus and for Chaim Weizmann. In 1941 he entered the United States; he chose San Francisco as his final residence in 1945 and became an American citizen in the following year. He died in 1953. Albert Einstein did not return from a trip to the United States in 1933; he became a fellow of the Institute for Advanced Study at Princeton. Even after World War II had come to an end, he never again set foot on German soil and would correspond only with those who had not succumbed to opportunism or collaboration. His search for a unified field theory was in vain, however, and his opposition to quantum mechanics and its Copenhagen interpretation drove him to the sidelines of science.

By concentrating here on the effects of this chauvinistic policy on the local community of scientists on the Telegraphenberg in Potsdam, however, we focus on the impact of the national policy within a smaller arena. Just as the political events in Germany and the rise of the Nazi Party can only be comprehended when we look at the accomplishments and mistakes of the competing political factions and the explosive potential that had been building up during the Weimar Republic (the Versailles Treaty, minority problems, impoverishment of the population through two serious economic crises, the splintering of the parties in the Weimar democracy, etc.), in the same way, the sudden reversal in research policy at Potsdam in 1933 can only be understood in the light of the already longstanding conflict between Freundlich and Ludendorff. (See Chap. 10.) Individuals whose ambitions had been frustrated for years now took the helm at both the national and the local level. Their actions were guided by a vengeful and rancorous principle: to do everything otherwise than under the hated Weimar Republic.

On the political position of Freundlich's counterpart Hans Ludendorff, the following can be ascertained: the combination of a family military tradition and his basically nationalistic mentality smoothed the way toward his supporting the Nazi takeover outright, along with the consequent politically motivated changes.[3] An obituary of 1941 notes: "His

order additional emphasis, the letter warned that a civil servant at the Geodesic Institute had been dismissed for refusing to use this salute. All civil servants, salaried employees, and wage earners under Ludendorff signed the circular as requested. Freundlich alone took another course. Rather than simply sign it, he appended the following text: "I would like to be informed about who was supposed to have issued this notice. To my knowledge none of the members of my institute have refused to execute any official instructions. [*Signed*:] Dr. E. Finlay Freundlich."[12] The reaction to Freundlich's unintimidated postscript was not long in coming. A direct personal attack had thus evidently been in the making. Undoubtedly Ludendorff had been the initiator of the circular, this time remaining skillfully behind the scenes. Two days later, on October 7, in a biting response Ludendorff refused to address the question about the originator of the notice, on the justification that he felt not in the least accountable to his subordinate. He informed Freundlich, furthermore, that a complaint had been raised against him through the director of the Geodesic Institute for "not returning the German salute to someone from his institute,"[13] and that he had passed this complaint on to the ministry, supposedly "at the request" of Vahlen.

Freundlich later recounted this sleazy train of events in an article entitled "The Hitler Salute Also Applies at the Einstein Tower" in the *Pariser Tageblatt* for March 25, 1934.[14] Ludendorff subsequently denied that this had been the cause of his dismissal, asserting that Freundlich's "placement in retirement (not 'dismissal') had been based on the so-called Aryan provision (§3 of the Law for the Restoration of the Professional Civil Service)" and not on his refusal to use the Hitler salute.[15] But this captious correction was mere pedantry, as Freundlich's sarcastic reply points out:

Newspapers have drawn conclusions that are not quite correct . . . from an exchange of insufficiently informed notes, which Potsdam with its highly developed sense of justice—which today many Germans are especially noted for, of course—attaches great importance to correcting. . . .

Understanding how such an essential distinction can exist between whether a scientist, to whom his country is indebted for a large research institute, had been dismissed after 23 years of service, as had actually been originally intended, or whether he had only been driven from his position and his country through the admitted use of denunciations (the world knows what the frequent outcome of these are in Germany) in order then to place him in retirement with vague prospects of a pension based upon so-called laws that cannot be justified ethically by any moral law in the world—thus, understanding how such an essential moral distinction can exist between these two courses of events certainly does demand an overly sophisticated sense of justice.[16]

In 1933 Freundlich had become intolerable in more than one way. First, not only was he himself of partly Jewish extraction, but he was also the spouse of a "full-blooded" Jew in a state that had assumed anti-Semitism as a central component of its ideology. Second, he was a protégé of Einstein and an active defender of relativity theory in a research climate in which the proponents of "Aryan physics" had just acceded to power, decrying that theory in particular as the perfect example of formalistic, shallow, and absurd "Jewish theory." Finally, he was constantly swimming against the stream, refusing to adopt the spirit of the *Führerprinzip*, which was beginning to pervade all academic institutions in Germany.

Thus like hundreds of other university teachers, Freundlich was forced to emigrate in 1933 as a result of the National Socialists' rise to power and subsequent persecution of Jews and dissenters.[17] Along with about a hundred other lecturers he chose as his new home not England, Russia, or the United States, but Turkey. The country had been turned into a republic in 1923 under Kemal Atatürk. In the spring of 1933 Atatürk initiated a reform of the universities, which received an additional boost from the massive influx of highly qualified political refugees from Germany.[18] Unlike the majority of emigrants, however, Freundlich seems to have left the possibility of returning to Germany open for some time. On October 3, 1933, shortly before his controversy with Ludendorff over the salute, he submitted an application for a two years' leave of absence from the institute to accept an appointment in Istanbul. In it he called the establishment of an observatory in the Mediterranean a necessity for "German astronomy's regaining its lead in astronomical research."[19] In addition he suggested that if interested the Ministry of Culture would be able to influence the organization of the observatory, "but only if the undersigned, who originated the plan, is allowed to retain some sort of official relationship with Germany." He added that only if his connections to his former institute were maintained could the "Institute of Solar Physics" expect further grants from Carl Bosch, for example (i.e., IG Farben). After arguing that the small institute depended on private contributors, Freundlich could not refrain from closing presumptuously:

I personally attach particular importance to my being guaranteed access to the research facilities at the Institute of Solar Physics in Potsdam—which institute I have built out of private funds, have planned, and have furnished with international scientific standing. Rights to these research facilities may well be taken from me but can never be denied.[20]

No other scientist, to my knowledge, tried to negotiate with the German science bureaucracy in such a self-confident, even demanding way after his de facto emigration. But he was naturally not the only one to miscalculate how to deal with the new rulers. The representatives of the democratic parties clearly underestimated the National Socialist Party and its subordinate organizations at the end of the Weimar Republic and the beginning of the Third Reich.

The ministry, of course, did not take Freundlich's unusual suggestion. He did not even receive a response to his application for a leave of absence,[21] and other measures were not taken against him probably only because he had already left the country. The files reveal that though no disciplinary proceedings were initiated against Freundlich, his salary was withheld from November 1, 1933, onwards and that he received no pension payments abroad.[22] In February 1934 Freundlich protested by letter to the Prussian Academy of Sciences, which had always been involved in the administration of his institute: "An attempt is being made to take the Institute of Solar Physics (Einstein Institute) away from me—which I had founded out of private funds and organized down to the last detail and then donated to the Prussian state—with the intention of robbing me of all my rights to it."[23] Freundlich's correspondence in this period indicates that he interpreted all these events as "machinations" of Ludendorff and his friends (Vahlen in particular) aimed specifically against him, and that he did not yet notice the systematic realignment of scientific institutions in general. In the sciences as in other academic fields, power changed hands in 1933: Einstein, a central scientific figure in Berlin, had become an exile in 1932, and in numerous interviews and statements he did not disguise his misgivings about the situation in Germany. Fritz Haber, director of the Kaiser Wilhelm Institute of Chemistry and an influential scientific policymaker in Berlin until 1933, was also forced to resign and emigrate. Walther Nernst, Max Planck, and Max von Laue were pushed aside. On the other hand, Philipp Lenard and Johannes Stark, promoters of the "Aryan physics" movement and avid supports of Hitler from 1924 on, were appointed to leading positions. Ludendorff and others like him, who had been smoldering for years in subordinate positions, could now finally implement long-cherished plans that had been barred by the

Fig. 49. Bust of Einstein (by Kurt Harald Isenstein, 1929) in the entrance hallway of the Einstein Tower, with a stone ("ein Stein") that is said to have stood symbolically in place of the bust during the Hitler years. (Photo by the author)

"Einstein clique." In short, the entire scientific policy of the Weimar Republic either broke down or at least was drastically altered.

Few scientific institutions could defy political alignment successfully. One apocryphal anecdote reflects the silent resistance of a few against unpopular instructions from above: when Ludendorff ordered the immediate removal of the bust of Einstein that had decorated the entrance hall of the Einstein Tower since its inauguration, one employee, Erich Strohbusch (1899–1975),[24] formerly a mechanic of Freundlich's, stored the bust in a safe place until the fall of the Hitler regime. It is just a charming myth, however, that he left a stone ("ein Stein") symbolically in its place. (See Fig. 49.)

In 1935 Paul ten Bruggencate (1901–61)[25] became Freundlich's successor as professor and chief observer at the Astrophysical Observatory and director of the recently renamed Institute of Solar Physics. Under his direction investigations on photospheric granulation were conducted at the Einstein Tower as well as photometric analyses of Fraunhofer lines. In 1941 ten Bruggencate moved to Göttingen, where another tower tele-scope had been erected before the end of the war with the support of the military,[26] and Harald von Klüber[27] took over direction of the institute in Potsdam. He redirected the observation program primarily to the mea-surement of magnetic fields in sunspots, which had some military rele-vance, because turbulence in the solar atmosphere is connected with radio interference on earth.[28]

But back to the fate of the exile Freundlich: in 1933 the 58-year-old had to start all over again in a land in which astronomy had lain fallow for centuries. The country's only observatory, constructed in 1577, had been destroyed in 1580 on the order of the superstitious Sultan Murat III.[29] Freundlich became professor of astronomy in the department of sciences at the University of Istanbul and in 1934 started to build up an astronomy department there with the help of his teaching assistant Wolfgang Gleiss-berg, who had previously held a position at Breslau University. Gleissberg later described the exhausting enterprise of creating a department out of nothing:

At the beginning there were absolutely none of the physical prerequisites for astronomical instruction or research: there was neither an astronomy department nor any kind of astronomical literature available, not to mention equipment or observational instruments. These deficiencies could be eliminated relatively quickly, thanks to the generous support the university received from the Turkish government. The cornerstone of the observatory building was laid in January 1935 in the university's park in the Beyazit district at one of the most elevated points of the city, and construction proceeded so briskly that we could furnish the offices already in the fall of the same year. In the following year the building was crowned with a Zeiss dome 6.5 meters in diameter, and a Zeiss astrograph (four-lensed, d = 30 cm, f = 150 cm) was mounted below it. The university provided adequate funds for the procurement of literature, and many colleagues abroad helped with donations of journal series and publications, so that in a short time a usable department library was available. Clocks and auxiliary instruments were also obtained.[30]

In fulfillment of his contract Freundlich wrote an astronomy textbook for publication in Turkish, and in this period a number of his articles ap-peared in professional journals, some in Turkish translation, some in French, and some even in German.[31] His most notable student was Paris Pismis, who later emigrated to Mexico and became the most prominent astronomer in that country. Despite rapid progress in the construction of the observatory building and the harmonious collaboration of the German professors with their Turkish teaching assistants and translators,[32] Freundlich decided to leave Istanbul in 1937 to accept a position as professor of astronomy in Prague.[33] There also he had an influence as an academic teacher on the next generation of astronomers: his student Zdeněk Kopal was later professor of Astronomy at the University of Manchester for some thirty years.[34]

The Nazi expansionism that eventually ignited World War II drove Freundlich away again in January 1939.[35] Freundlich fled Prague to accept an appointment at the University of St. Andrews in Scotland, where he stayed until 1959, once again building a new astronomy department and equipping an observatory.[36] One of Freundlich's sisters, Mrs. Elisabeth F. Wardale, accompanied him to St. Andrews. His brother Herbert and another sibling had already been living in England since 1933.[37]

World War II delayed construction of the observatory and department buildings at St. Andrews, however. From 1939, Freundlich read undergraduate astronomy as Napier Lecturer and gave courses on navigation for Royal Air Force cadets. His pamphlet *Air Navigation* barely missed fulfilling its original purpose of serving as a basic guide for pilots during the war, for it was published at the late date of 1945.[38]

Only after the Allies had forced the complete capitulation of Hitler's Germany in 1945 could Freundlich again consider building up his new institute and observatory in Scotland—his third already! Nevertheless, at age 60 this prospect seemed more attractive to him than returning to his native country, as the following excerpt from a letter to Mendelsohn shows, which simultaneously throws an interesting light on the fate of the Einstein Tower during the Nazi regime:

I receive many letters from Germany. The Einstein-Observatory is again operating as such, under von Klüber; directorship with Grotrian and H. Müller. They have apparently not been disturbed by the Russians, a) because I never excluded Russians in the years from 1920/33 and b) because Klüber in fact had been able to keep Hitlerism from the Institute.

They [once] offered me to return; this, however, is out of question, although Kellermann writes to me: "there is culture here in abundance." Otherwise also there is "more intellectual life than one would have thought possible after this horrendous collapse." . . .

Scotland has definitely nothing to offer in this respect. I am therefore building up my own little "island-world." . . .

I have engaged lately four collaborators of various sorts to join in very new research work of constructing a new type of telescope; about 6000 £ for this purpose have been granted to me generously. In a few months I am requesting Baruch Birnheim to give my Department a research-Student: a British and a Czech Astronomer are expected too who wish to join in the Telescope problem.

So, if condemned to Solitary confinement in this bleak Highland "Camp," I at least intend to make the "cell" a bit colourful.[39]

For the observatory's optics Freundlich chose a new Schmidt-Cassegrain telescope with a 36-inch aperture. This type of telescope, designed by the Hamburg optics specialist Bernhard Schmidt (1879–1935) in 1932, compensates for the aberrations of its spherical mirror by means of a correction plate within the instrument that intercepts the light's path to the mirror.[40] An 18-inch prototype was first constructed at the St. Andrews workshop in 1949 and tested at Mills Observatory, Dundee, before being installed at St. Andrews in 1951. The trial model was so promising that work was started in 1950 to construct the larger 36-inch version.

On January 1, 1951, Freundlich became the first Napier Professor of Astronomy and accordingly began to supervise doctoral students, notably L. T. Searle, W. Nicholson, Eric Gray Forbes, and Alan H. Batten.[41] A description by Batten, one of his most successful students, indicates the strengths and weaknesses of Freundlich's teaching style and gives an idea of how he dealt with his coworkers and students on a personal level:

Freundlich was a stimulating teacher who expected his students to stretch their minds. He was impatient with authority, especially if he felt it to have been exercised arbitrarily, and perhaps sometimes hasty in his decisions. These qualities, on occasion, made him a difficult colleague. His students saw another side of him, however, unless (indeed) those same characteristics were used on their behalf. . . . He was never quite at home in the British university system, which seemed to him to place too much emphasis on examination result and to lead to spoon-feeding. . . .

A few days before sitting my own final examinations, I met him at the observatory. No doubt with the intention of encouraging me, he remarked that he did not agree with examinations and believed that the decision should be made on a student's record. "But," he added, "you still have the chance to spoil your record!" He proceeded down the stairs, thoroughly enjoying his little joke—rather more than I did at that particular time. A week or ten days later, when I had not spoiled my record, his delight was obvious.[42]

Batten's depiction of Freundlich as a difficult colleague immediately calls to mind his confrontational relations with Ludendorff at Potsdam: it comes as no surprise that his relations with his successor in Scotland soured as well.

After Freundlich's retirement in 1959, D. W. N. Stibbs denied him access to the St. Andrews Observatory. Thus the man who had orchestrated the building of the new telescope was unable to take part in the final phases of its installation.[43]

CHAPTER 12

A Solitary Fate:
Photon–Photon Interaction

In 1930 Freundlich began to have serious doubts about the validity of the general theory of relativity in fully describing the propagation of light in the proximity of large masses like the sun. Analysis of observations from the solar eclipse of 1929 and reanalysis of data from the Lick Observatory expedition of 1922 and the Greenwich data of 1919 yielded an unexpectedly large light deflection value. His suspicion was strengthened by the equally stubborn problem of detecting gravitational redshift in the solar spectrum as predicted by Einstein. In the 1930's Freundlich began to surmise that the field equations of the general theory of relativity were only approximately correct and that light did not propagate along the geodesic lines of space-time but rather was forced by an additional interaction onto slightly different paths. Thus in the 1950's Freundlich's theoretical efforts terminated in his hypothesis of photon-photon interaction, or the reciprocal influence of light quanta upon each other. The natural assumption to make was that the effect of this scattering is proportionate to radiation density and to distance l in a radiation field through which a quantum of light is traveling. Applying the Stefan-Boltzmann law, which connects radiation density to the fourth power of temperature T, Freundlich set the redshift as:[1]

$$\frac{\Delta \nu}{\nu} = -A \cdot T^4 \cdot l \,. \tag{8}$$

He determined the constant $A \cdot 2 \cdot 10^{-29}$ degr^{-4} cm^{-1} from the available fixed star statistics, under the simplifying assumption that the average surface temperature of a fixed star class could be applied for T, and the average radius for l, thus obtaining the average observed redshift for the fixed stars of the relevant class. He was encouraged by the fact that he had obtained approximately concurring orders of magnitude for constant A. It even seemed feasible that an extension of the above law (Eq. 8) could explain cosmological redshift, because the dependency of redshift on distance was automatically built into the equation, while the resulting temperature approximated 2° K in order to agree with the Hubble constant estimate then accepted. He hoped, furthermore, to be able to explain the mysterious center-limb shift in the solar spectrum, because for purely geometric reasons a much larger photon-photon effect should occur for rays originating from the solar limb than for central ones at each given level in the solar atmosphere.

Yet even this final attempt to gather the many problems within a single unifying assessment failed. Objections were raised from disparate quarters against his arguments—in part against the data he used, in part against other effects implied by his formula for which there was absolutely no evidence.[2] New, fierce controversies with E. Margaret and G. R. Burbidge, W. H. McCrea, D. ter Haar, Max Born, and others isolated Freundlich once again within his field, this time permanently.[3] By this time Einstein had lost all interest in experimental tests of his theory. In 1952, when Born informed him of Freundlich's latest results on light deflection and gravitational redshift,[4] Einstein replied:

Verification of the theory is unfortunately much too complicated for me. We're all just poor blighters, you know! But Freundlich doesn't impress me one bit. If absolutely no light deflection, no perihelion motion, and no line shifts were known, the gravitation equations would still be convincing, because they avoid the inertial system (this ghost, which influences everything, but to which things in turn do not react). It really is odd that people are usually deaf to the strongest arguments, while they are constantly inclined to overrate precision in measurement.[5]

How different from the Einstein of 1911, who had pressed Freundlich so much to do everything within his means to test his predictions! (See Chap. 2.)

Freundlich, on the other hand, participated in an expedition as late as 1954 to observe a solar eclipse in Sweden. He was accompanied by two members of the staff of the Potsdam Astrophysical Observatory using instruments owned by the East German Academy of Sciences in Berlin.[6]

In a professional opinion to Joseph Naas, the director of this academy, Einstein aired his doubts about "whether the gains made by a new determination [of the deflection of light] justify the considerable costs of such research," since no significant improvement in precision of measurement over earlier observations could be expected with the available instruments.[7] Freundlich's project received support nevertheless, but unfortunately this opportunity to test light deflection anew also fell through because of cloud conditions during the eclipse. Thus Freundlich was denied this last opportunity to obtain his own data to investigate whether the deflection of light in the sun's gravitational field is in fact larger than Einstein predicted. Freundlich was not able to witness later technological improvements in measuring light deflection in a gravitational field, especially in the area of radio interferometry, nor the appearance of new quantitative tests of general relativity in the space age—for example, through the time retardation connected with light deflection. He accepted an honorary professorship in 1957 at the Johannes Gutenberg University in Mainz and spent his last days in Wiesbaden together with his wife, Käte, who survived him by many years. Freundlich died in Mainz on July 24, 1964.

Throughout his career Freundlich adopted and actively defended opinions that were at variance with the majority view, unswayed by the damage they might do to his professional reputation. At first, as an early supporter of Einstein, actually even as his first apologist among the German astronomers, he became a scapegoat for the likes of von Seeliger, conservative, antirelativist astronomers who sought an outlet for their opposition to the revolutionary theorist. Then, as the scientific community began increasingly to acknowledge and even to admire Einstein's relativity theories, Freundlich turned about-face and began to express his doubts publicly about the accuracy of the theory. When experiments on light deflection and gravitational redshift were being applauded everywhere as a triumph for the theory, Freundlich again took a view contrary to that of most physicists, astrophysicists, and astronomers. Twice his reputation was damaged to such an extent that few professional opportunities remained open to him. His relatively secure tenure as director of the Einstein Institute between 1921 and 1933 should be looked upon as an enclave won through bitter struggle within an otherwise hostile environment. Even his posts in exile in Istanbul and Prague, just as later at St. Andrews, were peripheral, professionally no less than geographically.

TABLE 6

Comparative overview of the results of various solar eclipse expeditions on light deflection at the sun's rim

Observatory (site)	Date of eclipse	Focal length (f) (cm)	Aperture (ϕ) (cm)	ϕ/f	Instrument (lens)	Field of plate	Number of plates	Exposure (sec)
Greenwich (Brazil)	May 29, 1919	570	10	1:57	Coelostat (double)	$2°.4 \times 2°.0$	7	28
		343	20	1:17	Coelostat (double)	2.7×2.7	16	5–10
Greenwich (Principe)	May 29, 1919	343	20	1:17	Coelostat (double)	2.7×2.7	2	2–20
Adelaide and Greenwich (Australia)	Sept. 21, 1922	160	7.5	1:21	Astrograph (quadruple)	7×8	2	20–30
Victoria (Australia)	Sept. 21, 1922	330	15	1:22	Astrograph (quadruple)	—	2	45
Lick I (Australia)	Sept. 21, 1922	450	12	1:37	Astrograph (double)	5×5	4	120–25
Lick II (Australia)	Sept. 21, 1922	150	10	1:15	Astrograph (quadruple)	15×15	6	60–102
Potsdam I (Sumatra)	May 9, 1929	850	20	1:42	Coelostat (double)	3×3	4	40–90
Potsdam II (Sumatra)	May 9, 1929	343	20	1:17	Astrograph (triple)	7.5×7.5	3	14–56
Sternberg (U.S.S.R.)	June 19, 1936	600	15	1:40	Astrograph (double)	3.5×3.5	2	25–35
Sendul (Japan)	June 19, 1936	500	20	1:25	Coelostat (double)	2.9×2.9	2	80
Yerkes I (Brazil)	May 20, 1947	609	15	1:40	Astrograph (triple)	4×4	1	185
Yerkes II (Sudan)	Feb. 25, 1952	609	15	1:40	Astrograph (triple)	4×4	2	60–90

SOURCE: von Klüber 1960.

[a] Distance of stars' position from the solar center, minimum and maximum, measured in units of solar radii.

[b] Photographic (ph) and visible (vis) magnitudes.

Thus the numerous breaks in the battered and tortuous course of Freundlich's career stemmed from his peculiar tendency not to conform. Nonetheless his contributions are unjustifiably forgotten today. As his student Batten observed:

There was certainly an element of paradox about his life. . . . He deserves his place in the history of attempts to test general relativity by observation, yet much of his time and energy near the end of his life was devoted to opposing the theory and exploring what now seems to have been a blind alley. . . . Although it now appears that he was wrong about general relativity, his skepticism of wide-ranging theories was in good scientific spirit, and it is the part of his legacy to me that I value the most.[8]

Limiting stellar magnitude[b]	Number of stars	Solar radii from center[a]		Check-field	Light deflection	Mean error	Reference
		r_{min}	r_{max}				
6 (ph)	7	2	6	No	1″.98	0″.16	Dyson et al. 1920
6 (ph)	11	2	6	No	0.93	—	
6 (ph)	5	2	6	No	1.61	0.40	
8.3	11–14	2	10	Yes	1.77	0.40	Dodwell and Davidson 1924
9.0	18	2	10	Not used	1.75 1.42 2.16	—	Chant and Young 1924
10.5 (ph)	62–85	2.1	14.5	Yes	1.72	0.15	Campbell and Trumpler 1923/24
10.4 (ph)	145	2.1	42	Yes	1.82	0.20	Campbell and Trumpler 1928
8.9	17–18	1.5	7.5	Yes	2.24	0.10	Freundlich et al. 1931a
9.5	84–135	4	15	Yes	—	—	Freundlich et al. 1933
9.6	16–29	.2	7.2	Not used	2.73	0.31	Mikhailov 1940
8.6 (vis)	8	4	7	No	2.13	1.15	Matukuma 1940
10.2	51	3.3	10.2	Not used	2.01	0.27	van Biesbroeck 1950
8.6	9–11	2.1	8.6	Yes	1.70	0.10	van Biesbroeck 1953

Freundlich's research exemplifies the difficulties involved in verifying a theory factually. The overview in Table 6 shows the great variation in the few values obtained for light deflection up to 1960. In general the epistemology of the time was mistaken in taking the relation between experiment and theory to be unproblematic: logical empiricists, for example, saw experimental results as the direct verification of theoretical predictions; and Freundlich seems to have agreed with this empirical viewpoint. He demonstrated his support for Hans Reichenbach's Berlin group of philosophers of science, for example, with a lecture in 1931 on the finiteness of the universe; and in 1951 he contributed an article on cosmology to the *International Encyclopedia of Unified Science* issued by the

Unity of Science movement led by Otto Neurath, Rudolf Carnap, and Charles W. Morris.[9] But simplistic demands for straightforward "verification" or Karl Popper's imperative of replacing it with an equally direct falsification of incorrect theories through observational data do not apply to the general theory of relativity. For decades on end, experimental measurements, instrumentation, and theories were being readjusted and refined to describe the very complex processes in and around the sun, tapping developments in other areas of technology, physics, and astrophysics such as radio interferometry, solar theory (solar atmosphere models), quantum mechanics (statistics), and solid state physics (Mössbauer effect).[10]

Only when this process came to an end in the 1960's and 1970's, six decades after Einstein's revolutionary publications, and only after Freundlich's death, did recent experimental technology permit a more or less satisfactory "verification" of the experimental predictions of the general theory of relativity.[11] Experiments previously held to be decisive proved on the contrary to be completely inadequate. An example for the special theory is the accumulation of mass by electrons, and for the general theory, light deflection and redshift measurements in the solar spectrum. Their accuracy had been overrated, while the systematic errors had been too large for the available technology to surmount. In retrospect this is also true of the experiments that the majority of Einstein's competent contemporaries had accepted as key verifications of relativity theory, such as Eddington's and Crommelin's measurements of light deflection in the sun's gravitational field of 1919 or St. John's results on gravitational redshift of 1923–24.[12] In this sense, Freundlich cannot be entirely blamed for his failure to produce a more precise test of general relativity. On the contrary, for the sake of accuracy he was relentless in searching out the weaknesses in his contemporaries' results. His ambitious scheme had just come a few decades too soon.

CHAPTER 13

A Bird's-Eye View

❖ The essential task of theory building here is not to codify
abstract regularities but to make thick description possible,
not to generalize across cases but to generalize within
them.
—Clifford Geertz (1973: 26)

Having come to the end of our tale let us move away from the historical
details and take a bird's-eye view of this case study. We have reconstructed
a well-documented struggle of one individual in an unfriendly environment, at the Berlin Royal Observatory under Struve, later at the Astrophysical Observatory under Ludendorff. We have also shown how, after
the failure of his first efforts, Freundlich's influential mentors made it
possible for him to continue his work at a different, newly established
institution, even providing him with his very own institute, the Einstein
Tower, at which to conduct his research—and that, too, at a cost of 1.5
million marks in a time of financial crisis.[1]

Taking up the metaphor suggested earlier, Freundlich's career can be
looked upon as the path of a test particle revealing the attractive and
repulsive potentials of his environment; Einstein on the other hand was
too renowned and otherwise too well protected to be deflected significantly by the pressures around him. As an ordinary, less conspicuous
figure, Freundlich was subjected to the full force of the various powerful
science cliques. Thus more comes to light about the different factions
within the Berlin scientific community, about the "line of defense" that
scientists formed around Einstein[2] against antirelativists, who were the
overwhelming majority not only at the Physikalisch-Technische Reichsanstalt, but also even within the Berlin astronomer community. The many
turns in Freundlich's journey through life demonstrate the considerable
deflective force of the repellent centers of power: until 1915 at the Berlin
Royal Observatory, from 1917 on at the Kaiser Wilhelm Institute of
Physical Research, and around 1922 in Potsdam in his daily struggles with

Ludendorff. It was only in 1933, with the Nazis' seizure of power, that a major turning point came not only for Freundlich and Mendelsohn but also for Einstein. Forced emigration meant an end to their collaboration, even though relations between Einstein and Freundlich had already become very strained in 1925 through differences of opinion and outlook.

It would be tempting to conclude from Freundlich's use of English in his letters to Mendelsohn in 1946 that he had become fully adjusted to his new life in Scotland.[3] However, the correspondence indicates otherwise. In February of that year he wrote: "One wishes me to become a British citizen; I feel quite happy being a Czech, longing for the cosmic citizenship to be created."[4] His positions at St. Andrews and Istanbul had been on the periphery both geographically and professionally. His disagreement with his successor Stibbs prevented him yet again from reaping the fruits of his labors in establishing the new observatory at St. Andrews. A similar case could be made for both Einstein and Mendelsohn. Each may have finally settled down in his adoptive country, but neither regained the creativity of the years before his exile. Einstein chose isolation at Princeton and made no significant scientific contributions after 1935; and Mendelsohn, according to Julius Posener, always felt like a stranger in Israel. He could realize only a few of his projects in Israel and the United States—the majority of his designs for the University of Jerusalem on Mount Scopus, for example, ran aground in the planning stage, and as a new immigrant to America he was not permitted to build anything at all during his first five years in the United States. Thus the Einstein Tower is the result of a rare combination of factors leading to the successful collaboration of these very different personalities at their most creative phase. The physicist's interest in the verification of his theory, the astronomer's ambitions to acquire a permanent workplace to conduct his researches, and the architect's dream of having the chance to design and build a structure according to his own plans converged on the fertile political ground in greater Berlin in 1920 to result in this unique building.

The distinction that comes out so clearly between the main actors and the secondary figures in a biography becomes blurred with added distance. We understand our hero, Freundlich, only when we are familiar with the complete network: with his enemies as well as with his friends. That is why Ludendorff has surreptitiously become a major character even though he was not directly affiliated with the Einstein Tower. Thus in writing the history of science we must devote much more attention to selected peripheral figures who do not appear in the compendia of "great scientists" and who are remembered today at best as specialists. Few cases are so well documented, but even where the documentation is lean, much can still be learned from a closer look at the careers of "common folks" in the world of science.

The next step is to reconstruct the equivalent in science to a history of everyday life: that is, a description of the daily routine in the laboratories of the Einstein Institute and the Astrophysical Observatory through an examination of their lab records. The lack of availability of this type of material was an unfortunate barrier in this case. The account of the daily battles between Freundlich and his superior Ludendorff was the closest substitute for this; however, a more in-depth look at the research conducted between 1924 and 1933 at the tower would have been desirable. (See Chap. 8.) Such a reconstruction may have brought us closer to understanding why the Einstein Tower did not ultimately fulfill its founding purpose of verifying gravitational redshift in the solar spectrum.

But comparing our case with others in the history of science, we see that neither the hand-to-hand skirmishes between the advocates and the opponents of a new theory nor the foundation of new institutions based on changing research interests that could not be incorporated within existing ones are peculiar to our case or to the 1920's. On the contrary, historiography that distinguishes between such periods of "revolutionary" change and "normal science" will find the events described in Chapters 4 and 10 typical examples of scientific revolutions, where the opposing sides come into direct conflict and accuse each other of dogmatism. Obviously, from this polar perspective all three of our main characters (Freundlich, Mendelsohn, and Einstein) stand decidedly on the side of the new trend. They do not lament about "crises" like Fritz Ringer's scholarly mandarins but seek tenable foundations for novel ideas. In this sense they are essentially different from the decadent turn-of-the-century *Weltuntergangs-Moderne* of Böcklin or Schiele, and rather comparable to the *Aufbau* modernity of Kandinsky or Klee.[5] This impression, founded on a comparison of their own attitudes, is also confirmed when looking at their reception. This search for a new framework, combined with the dissolution of old prejudices, is either emphatically underscored or critically admonished.[6]

The following new conditions fostered this unique collaboration:

The interplay between the various institutions at the same location, in this case the Berlin environs, is very much more complex when compared against earlier episodes in the history of science. These scientific institutions included the venerable Prussian Academy of Sciences in interaction with the new Kaiser Wilhelm Society and its individual institutes, and the traditional Berlin Royal Observatory together with the more recently founded Astrophysical Observatory[7] and the Einstein Institute.[8] There is a similar pattern in the United States. Shortly after the turn of the century the Carnegie Institution was established as a private organization with the legal status of a registered association; the Kaiser Wilhelm Society was founded in 1911 following this important model.[9] There were, in addition, the German organizations promoting the advancement of science: the Notgemeinschaft der Deutschen Wissenschaft and the Deutsche Forschungsgemeinschaft. Large firms (e.g., Zeiss or Badische Anilin- und Sodafabrik, or Siemens and Halske), each with its own research departments and interests, ought also not to be forgotten, as well as business organizations promoting scientific research, foundations like the Robert-Bosch-Stiftung or the Jubiläumsstiftung der Deutschen Industrie zur Förderung der Technischen Wissenschaften.[10] This point is along the lines of Thomas Hughes's idea of "the emergence of modern forms: networks, grids and interconnections."[11] The Einstein Tower is the architectonic result of an interlocking of these institutions' interests with those of individuals (such as Freundlich, Planck, and Einstein). None of the nodes in this network of interests could have realized this project on its own.

The Einstein Tower is an early example of a "modern," large-scale research establishment, for which mixed financing and, increasingly after World War II, multinationality are typical. The increasing dependency of basic research on large-scale scientific equipment has also been considered a typical feature of modernization in twentieth-century science, with an exponential rise in the cost of instrumentation.[12] Einstein, Mendelsohn, and Freundlich were equally convinced of the supranationality of science and culture.[13] The purely patriotic aspect of this endeavor proved to be one of its most serious weaknesses. The Einstein Tower was perpetually trying to catch up with its role model, the Mt. Wilson Observatory, and ultimately never succeeded. The comparative measurement of spectral lines in cooperation with Mt. Wilson, initiated in 1926, was a tardy and unsuccessful attempt at removing the national barrier.

A glance at the Einstein Tower's financing brings another revelation: the role of private contributors was of overwhelming importance, amounting to a million reichsmarks; thus they assumed most of the construction expenses and the purchase of crucial instruments, as well as operational costs. In order to collect such a large amount of money from private sources it was necessary to conduct an unprecedented publicity campaign (directed

Mitteilung an die Stifter der Einstein=Stiftung und Aufforderung zur Bildung einer „Freien Vereinigung der Freunde des Einstein=Observatoriums.

Am 6. Dezember des verflossenen Jahres hat in den Räumen des Turmteleskops der Einstein=Stiftung die Eröffnungssitzung des Kuratoriums der Stiftung mit anschließender Besichtigung der in Potsdam neu entstandenen Anlage stattgefunden. Fünf Jahre sind seit der Begründung der Stiftung verflossen, und mancher Stifter, der vor fünf Jahren seinen Beitrag stiftete, wird dieses Unternehmen vielleicht ganz aus dem Auge verloren haben.

Diese fünf Jahre sind ganz mit dem Entwurf und Bau der neuen Forschungs= stätte ausgefüllt gewesen, die sich der astronomischen Prüfung der Relativitäts= Theorie und den vielen ihr angrenzenden Problemen zu widmen gedenkt. Erst jetzt ist das Unternehmen so weit gediehen, daß den Stiftern nicht allein Rechnung über die Verwendung der Stiftungsmittel ab= gelegt werden, sondern ihnen auch das Geschaffene vor die Augen geführt und über die künftigen Aufgaben des neuen Institutes berichtet werden kann.

Das hier wiedergegebene Bild gibt den äußeren Anblick des von dem Architekten Erich Mendelsohn entworfenen Baues des Turmteleskops wieder. Eine ausführliche Beschreibung der neuen Anlage wird den Stiftern überreicht werden, sobald das große Teleskop selbst in allen Teilen fertiggestellt und erprobt sein wird.

Nur soviel mag schon an dieser Stelle mitgeteilt werden, daß in dem von einer Kuppel gekrönten Turm, von diesem vollkommen isoliert, noch ein zweiter Turm bis in die Kuppel hineinragt, in der sich ein System großer Spiegel und eine Linse von etwa 15 Meter Brennweite befinden. So wird in einer Höhe von etwa 20 Metern über dem Erdboden das Licht der Gestirne aufgefangen und lotrecht nach unten in ein unterirdisches Laboratorium geworfen, wo es durch große Prismen und andere das Licht zerlegende Apparate in Spektren aufgelöst werden kann, die im Farbbereich vom Roten zum Violetten eine Länge von 4 bis 12 Metern erreichen. In dem Laboratorium selbst werden in einem Spektralofen bei Temperaturen von 3000 und mehr Grad die chemischen Elemente in gleicher Weise zum Leuchten gebracht wie auf den Oberflächen der Gestirne und so vergleichende Studien über das Wesen ihres Leuchtens getrieben. Die Arbeiten im Laboratorium sind schon seit

Fig. 50. Notice to Einstein Foundation donors, 1925, printer's proof with mar-ginal notes. (From Main Archive of the Academy of Sciences, Berlin, Astrophysi-cal Observatory holdings, file no. 147, sheet 100)

primarily by Freundlich), with repeated appeals for contributions, frequent articles in the press, and widespread distribution of promotional photos.[14] Though this campaign fulfilled its purpose, it also aroused resentment among more traditionally minded scientists, who came to regard such publicity as part of a new trend toward sensationalism. This confrontation between modernists and antimodernists[15] culminated in 1933 when the latter were able to implement a complete reversal in scientific policy.

How successful the Einstein Tower's publicity campaign was is reflected in the fact that even publications like the chauvinistic newspaper *Germania* reported positively on the tower, although they tended to give preference to authors opposed to relativity theory.[16] Of course, the stylistic ambiguity of the tower must have contributed significantly to this, with its traces of "antimodernity" (in Herbert Mehrtens's sense) within avant-garde design.[17]

The public reception of the Einstein Tower clearly parallels that of relativity theory itself. The construction and inauguration of the observatory, as well as the projects of its staff (such as, in particular, the solar eclipse expeditions), came into the limelight in 1919 as at no other time; the same applies to the scientific theory. Erich Mendelsohn's extraordinary, even spectacular exterior design was a significant if not indispensable aid in attaining the necessary level of public attention. Prominent photographers were drawn to the architectural masterpiece, including Sasha Stone,[18] whose photos record a visit to the tower in 1928. The tower offered to the fashionable society of the Weimar Republic a more accessible, "artistic" surrogate for Einstein's notoriously incomprehensible relativity theory. Had the tower been adorned with an oversized telegraph mast of the likes of the 150-foot tower telescope at Mt. Wilson, it would not have appeared on the cover of the *Berliner Illustrierte Zeitung* (see Fig. 19) and would certainly not have been considered worth revisiting in 1928 (see, Fig. 51): "What could have well become a sobre, vapid, chimney-like structure became in his [Mendelsohn's] design a most interesting and in that time quite revolutionary building. With its curvilinear skyline, but nevertheless fitting so well to its task and to the surrounding landscape, it pointed far into future architectural developments."[19]

The Einstein Tower was considered unanimously by contemporaries as a symbol of the "new age," whether they liked it or not. It was quite in contrast to the accepted, retrospective styles, as seen in the Moorish style embellishing the Einstein Tower's immediate neighbor on the Tele-

Berliner Illustrirte Zeitung   Jg 37 (1928)   2195

DAS GROSSE TURM-TELESKOP in POTSDAM

Ein Besuch im Einstein-Turm

Fig. 51. The Einstein Tower in the *Berliner Illustrierte Zeitung* 37, no. 51 (1928). (From Freundlich 1928d [photos by Sasha Stone], by permission of Ullstein Bilderdienst)

graphenberg, the Astrophysical Observatory. Thus both Einstein's relativity theory and the tower telescope bearing his name were generally perceived as breaking with tradition. As is typical in such periods of revolutionary change, the subcutaneous historical continuities remained unnoticed, though. Here is how one contemporary put it: "Just as Einstein's discovery is distinguished by an abrupt break with the previously feasible scientific research, his architect also has used new construction methods that have hardly anything in common with former practices."[20]

If it is permissible to describe Berlin science of the 1920's as a subset of the contemporary German culture, then this study should have exposed the warp and woof of this cultural fabric, composed of intertwined personal, institutional, and cognitive strands, which themselves were spun from the peripheral elements of society, politics, architecture, and history. To close with the words of Erich Mendelsohn: "Do not forget that the individual creative work can only be understood when taken together with all the contemporaneous developments. The former is as dependent on the relativity of the latter's circumstances as the present and future are dependent on the relativity of history."[21]

Reference Matter

Abbreviations in the Notes

The following archives will be referred to in the notes in abbreviated form, followed by a call number:

AAW   Main Archive of the Academy of Sciences, Berlin, of the former German Democratic Republic, especially files of the Berlin Royal Observatory since 1913 at Babelsberg, and of the Astrophysical Observatory in Potsdam

ALO   Mary Lea Shane Archives of the Lick Observatory, Santa Cruz

AMPG   Archive of the Max Planck Society, Dahlem, Berlin

ASP   Archive for Scientific Philosophy in the Twentieth Century, Pittsburgh

BAP   Federal Archive (Bundesarchiv), Potsdam Section

BLC   Bancroft Library, University of California, Berkeley, Emil Fischer Papers

CPAE   Collected Papers of Albert Einstein, Boston University (editorial duplicate archive) and Hebrew University of Jerusalem (original archive)

DAM   German Museum of Architecture (Deutsches Architektur-Museum), Frankfurt

ETH   Library of the Federal Polytechnic (Eidgenössisch-Technische Hochschule), Zurich, especially Hermann Weyl Papers

GSA   Prussian Cultural Holdings, Secret State Archives (Geheimes Staatsarchiv Preußischer Kulturbesitz), Berlin

HUBU   Humboldt University in Berlin, University Archive, personnel file of Prof. Erwin Finlay Freundlich

JGUM   Johannes Gutenberg University, Mainz, Archive

KB   State Museums of Prussian Cultural Holdings, Art Library, Berlin (Staatliche Museen zu Berlin, Preußischer Kulturbesitz, Kunstbibliothek), Mendelsohn Papers

RAS	Royal Astronomical Society, London
RS	Royal Society, London, especially Larmor Papers
UAG	University Archives, Göttingen
UBG	University Library, Göttingen, Manuscripts Department

Notes

Notes to Introduction

1. For summaries of German astrophysical research conducted during the National Socialist period, see Kuiper 1946, Kiepenheuer 1946/48; cf. also Wolf-schmidt 1992/93, 1994 on the construction of new solar observatories after the outbreak of World War II.

2. For a general description of Weimar culture see, e.g., Gay 1968; on the rise of antidemocratic tendencies during the Weimar Republic see, e.g., Herf 1984, Sontheimer 1968.

3. See Dyson et al. 1920, Earman and Glymour 1980b.

4. See, e.g., Hentschel 1990a.

5. See Vierhaus and vom Brocke 1990.

6. The Zeiss Company in Jena was split into an eastern and a western branch after World War II; for a company history up to 1946 see Schomerus 1952; on the subsequent period see also Hermann 1989. Apparently, none of Freundlich's correspondence with Zeiss prior to 1928 has survived in the company archive.

7. See, e.g., Achenbach 1987; Banham 1954; Beyer, ed., 1961; von Eckhardt 1960; Köster 1974; Mendelsohn 1930; Whittick 1940/56; Zevi, ed., 1963, 1985; as well as Sharp 1966 and Pehnt 1981 for overviews on expressionistic architecture. For studies on the Einstein Tower within architectural history, see J. Krausse's contribution in Eggers, ed., 1995; James 1994.

8. The originals are now archived at the Pierpont Morgan Library, New York; see *International Herald Tribune* (Mar. 26, 1992): 20. The Freundlich–Einstein correspondence up to 1914 appears in vol. 5 of *The collected papers of Albert Einstein*: see Klein et al., eds., 1993c.

Notes to Chapter 1

1. On Freundlich see, e.g., the curriculum vitae attached to Freundlich 1911; as well as Freundlich 1930e; Kopal 1964; von Klüber 1965; Forbes 1972;

Pyenson 1974 (313–37), 1985 (228–38); Batten 1985. For personal information on Freundlich, I am indebted to his nephew Mr. Winfried F. Freundlich of Wiesbaden. For portrait photographs, see Fig. 4.

2. See Reitstätter 1954.

3. The diploma issued to Freundlich for his transfer to Leipzig lists the courses he attended at Göttingen in his first two terms there. These included: "Kant and Neo-Kantian Philosophy" (Edmund Husserl); "Theory of Instruments in Astronomy, with practice sessions" (Leopold Ambronn); "Differential and Integral Calculus" and "Differential Derivations" (Carl Runge); "Projective Geometry" and "Function Theory" (Felix Klein); "General Astronomy" (Karl Schwarzschild); "Practical Sessions in Physics for Mathematicians" (Waldemar Voigt); and algebra (Hermann Minkowski). UAG, section "Abgangszeugnisse" no. 4/7 Frl. 3377 Gal.

4. For more on Klein see, e.g., Tobies 1981, 1994; Rowe 1989; and references there. On Schwarzschild, see below, Chap. 5 n. 4.

5. See the record in the graduation file at UAG: "Phil[osophische] Fak[ultät], Prom. Spec., F," vol. II, 1909–1914. The astronomer J. Hartmann had just replaced Schwarzschild, who had accepted the directorship of the Astrophysical Observatory in Potsdam in 1909.

6. Struve, who came from a family of astronomers, succeeded Foerster as director of the Berlin Royal Observatory in 1904. For more on Struve see, e.g., Courvoisier 1920b, Dyson 1921. For a history of the Royal Observatory in Berlin, which moved to Babelsberg (a suburb of Potsdam near Berlin) in 1913 and lost the designation "Royal" in 1918, see Fritze 1976; see also Struve 1919, Prager 1926, Müller 1991.

7. See Batten 1985: 33.

8. See, e.g., Pyenson 1974, 1985, where particular attention is devoted to relativity theory; as well as Rowe 1989 and Sigurdsson 1991 on the Göttingen approach to "mixed mathematics."

9. This information on Freundlich's early development is based in part on a curriculum vitae on file at the Main Archive of the Academy of Sciences (AAW, Babelsberg Observatory, file no. 64, sheets 1, 2), and also in part on a résumé attached to his dissertation of 1910: Freundlich 1911.

10. Demonstrator at the Institute for Cosmological Physics at the German University in Prague, where Einstein had also been teaching since 1911.

11. Einstein 1911, which laid out what the author later referred to as his Prague theory.

12. Pollak to Freundlich, Aug. 24, 1911, CPAE no. 11 181. See also Clark 1971: 161–62.

13. Pollak to Freundlich, Aug. 24, 1911, CPAE no. 11 181.

14. For more information on the experimental consequences of his theory, see among others Forbes 1963; Earman and Glymour 1980a, b; Hentschel 1992a,

1993a; and references there; as well as, e.g., Freundlich's overviews 1916a, 1919b, 1955a.

15. Light would thus appear not to travel in a geometrically straight line. For astronomical orders of magnitude the physicist must define the path of light rays as "straight" lines, or the shortest distance between points A and B.

16. Cf., e.g., Einstein 1906, 1907a.

17. See also Einstein 1911, 1912, 1913, 1914, 1915a–c, 1915/25, 1916b, 1917, 1920; as well as Freundlich 1917, and below, Chap. 3, for a brief survey of the development of the general theory of relativity.

18. See, e.g., Einstein 1907c (454–59), 1911a (905); see also Hentschel 1992a, b.

19. Pyenson (1974) points out that Freundlich's education at Göttingen places him alongside Minkowski, F. Klein, and Hilbert, who all showed an interest in Einstein's theories: see, e.g., Tobies 1994. Unlike the Göttingen mathematicians, however, Freundlich constantly stressed the experimental aspects in his representations of relativity theory.

20. AAW, Babelsberg Observatory, file no. 64, sheet 2.

21. Freundlich to Campbell, Nov. 25, 1911, ALO (original English). On Campbell, see also Crelinsten 1983 and below, Chap. 2 n. 4.

22. Courvoisier had been observer at the Berlin Royal Observatory since 1905, becoming chief observer there in 1914. On Courvoisier see Poggendorff, *Biografisch-Literarisches Handwörterbuch* (Berlin: Barth, 1926–55) vols. 5–7a.

23. See Freundlich 1913d, 1916i.

24. See Courvoisier 1905, 1913, 1919, 1920a, 1932; Hopmann 1923; Kienle 1924.

25. The Notgemeinschaft der Deutschen Wissenschaft supported Courvoisier's research after World War I: see Richter 1971 (500), 1972 (42).

26. Freundlich's results for the cosmic medium incorporated the "smallest" possible density "determined by known relations" of 1/10,000 of average air density and an extremely high absorption of light over longer distances. The circumsolar medium's pressure on the face of the earth came to 46 atmospheres per cm², and he arrived at a lower limit of $18.4 \cdot 10^{32}$ g for the circumsolar medium's total mass. See Freundlich's appendix in Courvoisier 1913: 70–75.

Notes to Chapter 2

1. Einstein to Freundlich, Sept. 1, 1911, CPAE no. 11 199 (Klein et al., eds., 1993c, doc. 281: 317).

2. Solar eclipse photos also include the background field of stars. The stars whose light rays pass closest to the sun exhibit most clearly the effect predicted by Einstein. This is discussed in detail below in Chap. 4.

3. Einstein to Freundlich, Jan. 8, 1912, CPAE no. 11 202 (Klein et al., eds.,

1993c, doc. 336: 387); see also the letter dated Sept. 1, 1911, CPAE no. 11 199 (ibid., doc. 281: 317).

4. On Campbell see, e.g., Dyson 1939, Crelinsten 1983.

5. Freundlich to W. W. Campbell, Nov. 25, 1911, ALO (original English), C. D. Perrine, from 1909 director of the Argentinian State Observatory, was inspired by Freundlich to undertake a solar eclipse expedition to Brazil in 1912, but it was unsuccessful because of rain; see Perrine 1923, Stachel 1986 (226). Crelinsten (1983: 12–16) discusses other inquiries by Freundlich.

6. Campbell to Freundlich, Mar. 13, 1912, ALO: "We have [read] Einstein's paper . . . , and we shall be glad to assist you as far as possible in testing the question."

7. See, e.g., Campbell to Freundlich, June 6, 1912, ALO: p. 2: "For an adequate treatment of the problem which you have in hand, plates should be taken with the sun central, and the cameras should be rated to a stellar rate instead of the solar rate, as was the case in all these eclipse plates."

8. In later expeditions comparison photographs of the same region of the sky were taken for this purpose with the same instruments before or after the eclipse. See, e.g., Littman and Willcox 1991, chap. 8.

9. See Freundlich 1913b as well as, e.g., Campbell to Freundlich, Mar. 6, 1913, ALO: p. 2: "Your experience with the eclipse plates is about what we had expected: not only is the sun's image near the edge of the plate, but the aberrations of the camera lens at the edge of the plate are unavoidable; and, further, the clock was regulated to follow the sun and not the stars."

10. Freundlich 1913b, col. 371; on Schwarzschild, see below, Chap. 5 n. 4.

11. Einstein to Freundlich, Oct. 27, 1912, CPAE no. 11 205-1 (Klein et al., eds., 1993c, doc. 420: 503).

12. See Freundlich 1913b, 1914a–d.

13. Einstein to Freundlich, mid-Aug. 1913, CPAE no. 11 203-1 (Klein et al., eds., 1993c, doc. 468: 550); see also the almost identical wording in a later letter of the same month, CPAE no. 11 204 (ibid., doc. 472: 554–55). The word *Eifer* (diligence, zeal) appears in almost all of Einstein's letters to Freundlich written in 1911 and afterwards.

14. Einstein to Freundlich, Jan. 8, 1912, CPAE 11 202. See also Einstein to Freundlich, Sept. 1, 1911, CPAE no. 11 199, and Sept. 21, 1911, CPAE no. 11 201-2 (Klein et al., eds., 1993c, docs. 281, 287: 317, 326); and Einstein's letter of Aug. 1913, in which he makes his own assessment of this problem: "Your plan to observe stars in the vicinity of the sun during the day interests me particularly. This ought to be possible if little particles of the order of magnitude of light waves were not suspended throughout the atmosphere bending the light just a bit. I fear your plan could fail on this. But you will be more familiar with these matters than I" (CPAE no. 11 204 [ibid., doc. 472: 554]). See also Hertzsprung to Schwarzschild, Jan. 16, 1916 (UBG, Schwarzschild Papers, folder 84/2): "It will be

difficult to test Einstein's theory on Jupiter. Only the satellites come into consideration practically. A few hundredths of a second of arc is the limit of what I can establish. But in the coming years Jupiter is positioned too low for such precision measurements at Potsdam."

15. See Einstein 1911: 904–5; cf. Forbes 1963, Earman and Glymour 1980a, Hentschel 1993b.

16. For contemporary discussions of pressure and Doppler shifts see, e.g., Humphreys 1908, Fabry and Buisson 1910; cf. Hentschel 1995b, chap. 4. On Julius's interpretation of these shifts see Hentschel 1991 and, e.g., Julius 1910.

17. Einstein to Freundlich, Jan. 8, 1912, CPAE no. 11 202-2 (Klein et al., eds., 1993c, doc. 336: 387–88).

18. See Forbes 1963; Earman and Glymour 1980a; Hentschel 1991, 1992, 1993a; and, e.g., St. John 1928.

19. Cf., e.g., Hentschel 1991.

20. Freundlich 1916a: 391–92.

Notes to Chapter 3

1. For a description of the course of development of these theories both historically and scientifically see, e.g., Norton's and Stachel's contributions in Howard and Stachel, eds., 1989, and the introduction to CPAE nos. 4, 6.

2. Cf. Einstein and Grossmann 1913.

3. Einstein to Freundlich, Oct. 27, 1912, CPAE no. 11 205-3, 4 (Klein et al., eds., 1993c, doc. 420: 504).

4. Einstein to Freundlich, Sept. 1, 1911, CPAE no. 11 199 (Klein et al., eds., 1993c, doc. 281: 317).

5. This can evidently be taken as a precursor to Popper's doctrine of falsification. See in this regard Einstein's article "Induktion und Deduktion in der Physik," *Berliner Tageblatt*, no. 617, Dec. 25, 1919, supplement 4.

6. Einstein to Freundlich, Aug. 1913, CPAE no. 11 204-2 (Klein et al., eds., 1993c, doc. 472: 555); see also Freundlich 1913c; de Sitter 1913a, b.

7. This proved to be premature: see the contributions by John Norton and John Stachel in Howard and Stachel, eds., 1989.

8. Einstein to Freundlich, Aug. 1913, CPAE no. 11 204-1 (Klein et al., eds., 1993c, doc. 472: 554).

9. See Hentschel 1992a, b.

10. See, e.g., Einstein to Schwarzschild, Jan. 9, 1916, CPAE no. 21 561-4, part 3, where he counters Schwarzschild's reservations about the measurability of light deflection off Jupiter's moons with the words "It has to work" ("Es muss gehen"). Einstein's reaction to the results of Eddington and Crommelin's May 1919 measurements published in Dyson et al. 1920 is the same: he sent off a number of triumphant letters on the very same day that he was notified in late

1919, among these one to his mother and one to Planck (CPAE no. 19 286), as well as a telegram to *Nature*. All this hardly agrees with the above myth—see Hentschel 1992b for detailed arguments against the Rosenthal-Schneider story.

11. At this time *professore straordinario* at the Milan Polytechnic. See, e.g., Abraham 1912; Einstein 1912c, d. On Abraham, see the contributions by Cattani and De Maria in Howard and Stachel, eds., 1989: 160—74.

12. This Finnish physicist was lecturer at the University of Helsinki from 1910 to 1918; see, e.g., Nordström 1912. For contemporary assessments, see Einstein and Fokker 1914, von Laue 1917; cf. also Isaksson 1985, North 1992.

13. Mie became regular professor of physics at Greifswald University in 1905 (until 1916); see, e.g., Mie 1912/13, 1914. See Abraham 1914, Einstein 1913b, contemporary survey articles on alternative theories; see also Pais 1982, chap. 13.

14. Einstein to Freundlich, Aug. 1913 (original emphasis), CPAE no. 11 203 (Klein et al., eds., 1993c, doc. 468: 550—51).

15. See Mie 1914.

16. Einstein to Freundlich, ca. Jan. 20, 1914, Klein et al., eds., 1993c, doc. 506: 594.

17. See, e.g., Freundlich to Campbell, Feb. 6, 1913, ALO, Campbell Papers, Freundlich folder.

18. See Campbell to Freundlich, Mar. 6, 1913, ALO: "If the Lick Observatory organizes an expedition to observe the total solar eclipse in Russia, August 1914, we shall plan to obtain photographs meeting the requirements of your problem; that is, with the solar image in the center of the field and with the driving clock adjusted to follow the stars. It is not certain that we shall send an expedition, but I hope that we may. . . . However, as we say in this country, it is well not to count the chickens before they are hatched; that is, we may not go to the eclipse, or if we go, clouds or other features may prevent the success of our plans." Campbell did succeed in getting funding for the expedition (see his telegram to Freundlich of May 27, 1913, and Freundlich's reply of May 28, 1913), but the expedition failed to bring home any usable exposures.

19. For details on the preparations for the expedition, see Kirsten and Treder, eds., 1979, vol. 1, chap. 4—in particular, Freundlich's application for funding: "Gesuch an die Akademie der Wissenschaften um Unterstützung für die Teilnahme an der Sonnenfinsternisexpedition auf die Krim" (164—66).

20. Einstein to Freundlich, ca. Jan. 20, 1914, Klein et al., eds., 1993c, doc. 506: 593—94.

21. See Freundlich's report to the Prussian Academy of Sciences of Oct. 7, 1914, AAW, II-VII-158, sheets 19/1—8; and Planck to Einstein, Mar. 19, 1918, CPAE no. 19 273-1: "Freundlich should only address the academy with a report describing the circumstances and in which he requests that the academy take steps with the relevant authorities to have the instruments confiscated in Odessa returned as soon as possible. I will then do all I can to support this petition."

22. See Dyson et al. 1920 for the original publication, transcriptions of scientific discussions, etc., in *Monthly Notices of the Royal Astronomical Society* 80 (1919/20): 96–118; *Observatory* 42 (1919/20): 119–22; as well as the *Proceedings of the Royal Society* (London) 97 (1920): 66–79; and Clark 1971: 228ff., Earman and Glymour 1980b, and Moyer 1979 on their reception. See also Freundlich's own popular article on the "victory" of Einstein's theory: Freundlich 1919a; see in addition below, Chap. 9, for Freundlich's subsequent criticism of Eddington's results.

23. See, e.g., Freundlich 1923b, Joel 1923, Ludendorff 1923/24, von Klüber 1926.

Notes to Chapter 4

1. Freundlich 1915/16a: 115.

2. Freundlich 1915/16b: col. 17.

3. Ibid. On statistical astronomy, cf. also Trümpler and Weaver 1953, Paul 1993.

4. Freundlich 1915/16b: col. 18. For later data cf., e.g., Trümpler and Weaver 1953: 291, 354, 566.

5. Einstein also assumed a static cosmos in the first cosmological application of his field equations in February 1917, until Friedman, Lemaître, and others arrived at the dynamic solutions to the field equations in the 1920's and 1930's— see, e.g., North 1965, and the contributions by Ellis and Kerszberg in Howard and Stachel, eds., 1989, as well as references there.

6. Gravitational redshifts can naturally also be given in Doppler shift units— then the relative motion necessary to yield the observed redshift is indicated in Doppler equivalence velocities.

7. Systematic errors could originate, e.g., from disregarding the differences between physical parameters of terrestrial light sources (i.e., pressure, temperature, or field influences) when comparing them against stellar atmospheres.

8. Freundlich examined in detail only spectral classes A and B, in his words "because the available observational data are not yet complete enough." See also Ludendorff 1916.

9. Freundlich 1915/16b: col. 21. Prominent helium lines characterize the spectra of class B stars.

10. Einstein to Sommerfeld, July 15, 1915; reprinted in Hermann, ed., 1968: 30.

11. See Freundlich 1915/16a: 117. Applying Freundlich's values, however, we get $f/f_\odot = 4.7/0.635$ and $\rho/\rho_\odot = 0.1$, resulting in $23.4 m_\odot$!

12. From 1882 professor of astronomy and director of the observatory at Munich University as well as president of the Astronomische Gesellschaft (Astronomical Society) from 1897 to 1921. On von Seeliger, see Auwers 1905, Eddington 1923, Kienle 1925, Paul 1993 (55–79, 145–49, 175–89, 237–45).

13. See AAW, Babelsberg Observatory, file no. 65, sheets 16, 17.

14. See in this regard Freundlich 1915a; see also von Seeliger 1906b, 1915; Pyenson 1974.

15. Freundlich to Struve, Aug. 7, 1915, AAW, Babelsberg Observatory, file no. 65, sheet 19.

16. Freundlich 1915/16b: col. 21.

17. Ibid.: cols. 21–22.   18. All quotes ibid.: col. 22.

19. Von Seeliger 1916: col. 84.   20. Ibid.

21. Ibid. Von Seeliger naturally could not know that developments in the decades subsequent to this debate would lead to the permissibility of much higher stellar densities.

22. Ibid.: col. 85.

23. Ibid.: col. 86.

24. Von Seeliger to Struve, Jan. 12, 1916, AAW, Babelsberg Observatory, file no. 65, sheets 41, 42; reprinted in Kirsten and Treder, eds., 1979, vol. 1: 171–72.

25. Von Seeliger to Struve, Aug. 15, 1915, AAW, Babelsberg Observatory, file no. 65, sheet 15.

26. Einstein to Sommerfeld, Dec. 9, 1915; in Hermann, ed., 1968: 37.

27. Schwarzschild to von Seeliger (undated: 1916), UBG, Schwarzschild Papers, no. 193.

28. This also applies to today's high-energy physics, where individual theoretical departments have the task of finding reasons why certain parameters (e.g., bonding strengths at specific energies) have one particular value and not another. Only by connecting phenomenological models to larger theoretical networks do they gain credibility.

29. Einstein 1916c: 123.

30. Handwritten parenthetical added by Einstein.

31. Einstein to Struve, Feb. 13, 1916, CPAE no. 22 299; or AAW, Babelsberg Observatory, file no. 65, sheet 54, reprinted in Kirsten and Treder, eds., 1979, vol. 1: 173.

32. See the following quote.

33. Einstein to Sommerfeld, Feb. 2, 1916 (pointed brackets indicate text deleted by Einstein), partly reprinted in Hermann, ed., 1968: 38–39; see also Pyenson 1974: 330–32.

34. Such as in January 1916 to Schwarzschild.

35. See n. 33 above (original emphasis).

36. Einstein later distanced himself similarly in 1920 from Hans Ludendorff, Freundlich's most influential opponent. On Ludendorff, see below, Chap. 10 n. 4.

37. Sommerfeld to Weyl, July 7, 1918, ETH no. HS 91: 751 (my emphasis); cf. also Sigurdsson 1991: 160.

38. See, e.g., Einstein 1917b: 90.

39. CPAE no. 9 263.

40. Einstein 1917b (90), 1917c (194–95); he then referred to a recent survey by Freundlich (1919b).

41. See, e.g., the documentation in North 1965, chaps. 5–7; Hentschel 1990a, sec. 1.4.

42. See, e.g., North 1965, chap. 7; Hetherington 1982.

43. See in this regard, e.g., Forbes 1963; Mattig 1995; Hentschel 1992a, 1993a, b, 1995b.

Notes to Chapter 5

1. See Kirsten and Treder, eds., 1979, vol. 1: 7ff., 95ff.; and Fritz Stern's contribution in Vierhaus and vom Brocke, eds., 1990: 516–51.

2. See, e.g., Kirsten and Treder, eds., 1979, vol. 1: 146–57; ibid., vol. 2: 67–83; Heisenberg 1971; Macrakis 1986.

3. This financier provided the Prussian Academy of Sciences with an endowment called the "Stiftungsfond der mathematisch-physikalischen Klasse"—see Kirsten and Treder, eds., 1979, vol. 1: 102–3, where Einstein's role in the Berlin scientific community from 1914 to 1933 is documented extensively.

4. Director of the Astrophysical Observatory in Potsdam from 1909 and member of the Prussian Academy of Sciences. On Schwarzschild, see, e.g., Einstein 1916d, Sommerfeld 1916, Hertzsprung 1917, Blumenthal 1918, Oppenheim 1923, ten Bruggencate 1955, Treder 1974, Diecke 1975, Struve 1975.

5. Einstein to Freundlich, Dec. 7, 1913, CPAE no. 11 206-1 (Klein et al., eds., 1993c, doc. 492: 581). Only nine months remained before the solar eclipse.

6. See the letter quoted below as well as Kirsten and Treder, eds., 1979, vol. 2, docs. A 307; B 9, 11.

7. Namely 3,000 marks each from Emil Fischer and from the magnate von Krupp von Bohlen und Halbach, who had made Freundlich's acquaintance through Fischer. See also Freundlich to Fischer, Dec. 22, 1913, Jan. 18, 1914, and Jan. 22, 1914, BLC, Fischer Papers.

8. On Einstein's attitude toward experiments in the period, see Hentschel 1992b.

9. Einstein to Freundlich, ca. Jan. 20, 1914 (original emphasis), CPAE no. 11 207-1 (Klein et al., eds., 1993c, doc. 506: 593–94).

10. Original emphasis. Cf., e.g., the facsimile of the first page of this letter in Eggers, ed., 1995: 40; see n. 5 above. On Struve see above, Chap. 1 n. 6.

11. See an opinion submitted by Struve to Naumann on Einstein's petition for a five-year research grant to enable Freundlich to test the general theory of relativity, reprinted in Kirsten and Treder, eds., 1979, vol. 1, doc. 91: 169–70.

12. The Berlin observatories were known for their assiduous data collection;

see, e.g., Hassenstein 1941; Herrmann 1975, 1981b. On the history of stellar spectroscopy see Hearnshaw 1986.

13. See docs. 90–96 in Kirsten and Treder, eds., 1979, vol. 1: 168–75.

14. See von Seeliger to Struve, Jan. 12, 1916, and Jan. 26, 1916, AAW, Babelsberg Observatory, file no. 65, sheets 44, 45.

15. On the new building, see Eggert 1914, Struve 1919.

16. Einstein to Schwarzschild, Jan. 9, 1916, CPAE no. 21 561-4, 5 (original emphasis).

17. The importance of Göttingen's intellectual environment (*Denkkollektive*, Fleck's definition) to Weyl, for instance, is demonstrated in Sigurdsson 1991.

18. Schwarzschild suffered from the advanced symptoms of a skin disease he had contracted during his military service through exposure to poison gas.

19. See, e.g., Einstein to Hilbert, Mar. 30, 1916, CPAE no. 13 097, in which the option of finding a job for Freundlich at Göttingen is discussed.

20. Einstein to Hugo Andres Krüss, *Referent* at the Prussian Ministry of Culture, in Ilse Einstein's hand, Jan. 10, 1918, Manuscripts Department of the Staatsbibliothek Berlin, Acta Preußische Staatsbibliothek, Kaiser Wilhelm Institutes folder XXVI, published in full and quoted from Castagnetti et al., eds., 1994: 106–7.

21. Freundlich's own applications to observatories outside Berlin (such as Heidelberg and Vienna) failed, and initial plans to habilitate were predictably dashed by his superior.

22. This contract is available at AMPG, Dept. I, Rep. 34, folder 2. On the background to this contract, see also Freundlich to Mendelsohn, July 1, 1917, pp. 3–4, KB; and Freundlich to Fischer, July 11, 1917, BLC.

23. During World War I Freundlich had been working for the Artillery Testing Commission on direction-finding techniques: AAW, Babelsberg Observatory, file no. 64, sheets 28–30; and Freundlich's correspondence with Mendelsohn in 1917, KB.

24. Planck to Einstein, Dec. 9, 1917, CPAE no. 19 262. On Planck's importance as a science administrator and spokesman for German science, see Heilbron 1986.

25. The accounts were balanced quarterly; see §4 of his contract of Feb. 1, 1918, AMPG, I, 34, Freundlich folder. Planck's letter is filed as CPAE no. 19 283.

26. See Freundlich to Mendelsohn, July 9, 1917, Mendelsohn Papers, KB; Einstein to Freundlich, ca. 1917 (undated), CPAE no. 11 223.

27. Freundlich to Felix Klein, Mar. 11, 1918, UBG, Nachlaß 22B. On the Potsdam Astrophysical Observatory, see below, Chap. 6 n. 3.

28. See the documents cited below, Chap. 11 nn. 19, 20 (and text thereto). Freundlich's plans date back to the beginning of 1917: see his letter to Mendelsohn of Jan. 18, 1917 (2, KB), on the "plan of a privately funded foundation for the purpose of allowing me to work here freely."

Notes to Chapter 6

1. Karl Friedrich Zöllner, a pioneer in this field with a particularly strong influence on Berlin astronomy, coined the term *Astrophysik* in 1866: see Herrmann 1975, 1981a. On the early history of astrophysics, see also Hearnshaw 1986; on the importance of amateurs in its early stages, see Lankford 1981, Krafft 1981.

2. See, e.g., Gingerich, ed. 1984. For a systematic comparison of German and American contributions, see Herrmann 1973.

3. On the history of the Potsdam Astrophysical Observatory see: Scheiner 1890; Hassenstein 1941; Ludendorff 1942; Gußmann 1975; Scholz 1975; Wempe 1975; Herrmann 1975, 1981b; Eggers, ed., 1995 (up to 1900): 11–25; see also Spieker 1879, Saal 1901, von Nostitz 1966, Knobloch and Weiss 1987, Müller 1991. On H. C. Vogel, see, e.g., Herrmann 1976, Hearnshaw 1992, Wolfschmidt 1991/92.

4. See Kirsten and Treder, eds., 1979, vol. 1, docs. 94, 95: 173–75.

5. In 1877 he became assistant to Hermann Vogel at the Astrophysical Observatory, where he eventually worked his way up to the directorship in 1917, which he held until 1920. He was valued by his colleagues for his "thoroughness, methodicalness, and conscientiousness": see Ludendorff 1925a: 163, 172; R. Müller 1925; Herrmann 1974.

6. Ludendorff 1925a: 173; see also id. 1925b: cols. 199–200.

7. Ludendorff 1925a (173): "Müller tried to allow his subordinates as much freedom as possible, because he was convinced that everyone works hardest when allowed to carry out tasks that he enjoys the most." In this respect he certainly was unlike Struve or Ludendorff. (See below, Chap. 10.)

8. See Kirsten and Treder, eds., 1979, vol. 1, docs. 104–6: 183–87.

9. Born in Pfaffendorf, near Koblenz, von Laue took his doctorate under the guidance of Planck in Berlin in 1903 and completed his postdoctoral thesis (Habilitationsschrift) in 1906 on the thermodynamics of interference phenomena. After a professorship in Zurich and Frankfurt, he became full professor of theoretical physics at Berlin University in 1919 and a member of the Prussian Academy of Sciences. On von Laue see, e.g., Ewald 1960, Franck 1960, Hermann 1973, Herneck 1979.

10. On Ludendorff, see below, Chap. 10.

11. AAW, II-XIV, 6, sheet 7d.

12. See Kirsten and Treder, eds., 1979, vol. 1, doc. 107 (187): transcript of a meeting at the Ministry of Culture on Jan. 10, 1922.

13. See Kirsten and Treder, eds., 1979, vol. 1, docs. 107, 116; vol. 2, docs. 374–76, 378–79, 414–16, 456, 467, 474, 476, 483, 502, 505.

14. See, e.g., Freundlich's articles on telescope development: Freundlich 1950c, 1953e.

15. On Hale, see, e.g., Adams 1938, Wright 1972, Hufbauer 1991: 73–79.

16. For photographs see, e.g., Müller 1991: 188–91.

17. On its history, see, e.g., Hale 1905, 1915; for photographs see also Müller 1991: 196–99.

18. See Lippmann 1895, Stoney 1896, Villiger 1926, Hale and Nicholson 1938: 3–7; on the strengths of coelostats, see also Hale 1905; and on the Snow telescope, Briggs 1991.

19. See, e.g., Hale 1908, 1915; Hale and Nicholson 1938: 5–15; and on later tower telescope development, e.g., Schmitz 1984: 291–95; see also Fig. 10 in this volume.

20. On Hale's and Adams's research see, e.g., Hufbauer 1991: 76–78; on St. John see Earman and Glymour 1980a, Hentschel 1993a, and references there.

21. Von Laue's comment in 1920 on a survey by the Würzburg physicist Ludwig C. Glaser on recent measurements of the Einstein effect in the spectrum, "Einstein-Effekt im Spektrum"; see Schwarzschild 1910.

22. Carl Bosch (1874–1940) was head of the chemical manufacturer BASF and later of the chemical trust IG Farben. He was an instrumental figure in forming science policy and providing industrial funding. Freundlich demonstrated his good will toward this power broker by designing and later monitoring the construction of a private observatory for Bosch.

23. Freundlich to Mendelsohn, Aug. 17, 1918, KB.

24. See Freundlich to Mendelsohn, Aug. 17, 1918, KB.

25. W. Nernst noted on a draft of the petition of Dec. 31, 1919: "The moment for this action is certainly opportune!"

26. AAW, Astrophysical Observatory, file no. 147: 2. Cf. Eggers, ed., 1995: 77.

27. "Aufruf zur Albert-Einstein-Spende," Kirsten and Treder, eds., 1979, vol. 1, doc. 98 (177), AAW, Astrophysical Observatory, file no. 147 (original emphasis).

28. See, e.g., the publication on the state of the appeal, "Albert-Einstein-Spende," AAW, Astrophysical Observatory, file no. 147, 1920; on contemporary research support, see Forman 1974.

29. See Kirsten and Treder, eds., 1979, vol. 2, docs. 367–68, 382, 393.

30. Freundlich for the Einstein Donation Fund to Minister of Culture Konrad Haenisch, May 5, 1920; reprinted in Kirsten and Treder, eds., 1979, vol. 1, doc. 100: 178.

31. See AAW, Astrophysical Observatory, file no. 147, balance sheet to 1921; ibid., no. 147, notice of the Einstein Donation Fund board, by E. F. Freundlich, 1922.

32. See Schlicker 1979: 434. The inflation rate in 1919 was 58.1 percent; in 1920, 113.1 percent; in 1921, 28.1 percent; in 1922, 1,024.6 percent; and in 1923, 105.8 million percent! Figures from *Brockhaus Enzyklopädie*, 19th ed. (Mannheim: F. A. Brockhaus, 1989), vol. 10: 492. See also Eulenburg 1920.

33. See Kirsten and Treder, eds., 1979, vol. 1, 180–81; vol. 2, docs. 364–65, 407.

Notes to Chapter 7

1. On Mendelsohn see, e.g., Mendelsohn 1930 (also reprinted); Whittick 1940/56; Banham 1954; von Eckhardt 1960; Zevi, ed., 1963, 1985; Achenbach 1987; Kühne 1994; Limberg and Staude 1994.

2. According to a letter from Käte Freundlich to Luise Mendelsohn, Apr. 29, 1970, Mendelsohn Archive; cited by Achenbach in Eggers, ed., 1995: 156 nn. 74, 86.

3. See, e.g., Freundlich to Mendelsohn, May 2, 1918, and May 16, 1918, KB. Both Freundlich and Mendelsohn were Bach enthusiasts; cf., e.g., Neutra, ed., 1986: 53.

4. Freundlich to Mendelsohn, Aug. 21, 1918, KB.

5. See the selection of Mendelsohn's letters by Beyer, ed., 1961: 40. For a complete collection of the preserved observatory sketches by Mendelsohn from 1917 and later, see the illustrations in Eggers, ed., 1995: 53–75.

6. Freundlich raised real, practical arguments against one of Mendelsohn's observatory designs that the latter had enclosed for Käte Freundlich. Cf. Freundlich to Mendelsohn, Aug. 25, 1917, with enclosure to K. Freundlich of Aug. 19, 1917, no. 114, Mendelsohn Archive; quoted by Achenbach in Eggers, ed., 1995: 58–59.

7. Mendelsohn to Freundlich, Aug. 31/Sept. 1, 1917, enclosed with letters to L. Mendelsohn of the same dates, no. 128, transcription in Mendelsohn Archive; quoted by Achenbach in Eggers, ed., 1995: 62.

8. Typed letter, Mendelsohn to Freundlich, Oct. 29/30, 1917, KB. Ditto marks replaced by words in square brackets.

9. See Achenbach 1987: 16, 52–60. On Mendelsohn's characteristic manner of sketching in quick, full strokes, his modeling and drafting, see also ibid.: 10; Limberg and Staude 1994: 4–5.

10. Freundlich to Mendelsohn, July 2, 1918, KB, Mendelsohn Papers; also quoted in Achenbach 1987: 61–62. See also Freundlich 1969.

11. See Achenbach 1987: 61, 63, fig. 21.

12. Freundlich to Mendelsohn, July 2, 1918, KB.

13. According to Hans G. Beck, a longtime staff member at the Zeiss optical company's astronomical apparatus development division in Jena, the vertical arrangement for the coelostat adopted in Potsdam offered "significant advantages over the American version" because of the way the mirrors were linked together (pers. comm., April 19, 1994): the entire mirror unit can be rotated around a vertical axis; cf. in particular von der Pahlen 1926a, and Staude in Limberg and Staude 1994: 82–83, which includes photographs of instrumentation currently

used in the tower. All the instrumentation below the redirecting mirror has been modernized since.

14. By 1929 there were six tower telescopes: two on Mt. Wilson, one each in Potsdam and in Arcetri (all refracting telescopes), one reflecting telescope in Pasadena, and a small one at the Heliophysical Institute, Utrecht; see, e.g., Abetti 1929: 62–71.

15. Among these, the young Richard Neutra; see N.N. 1963.

16. See Pehnt 1981: 117; cf. also Mendelsohn 1930; Schmidt 1930; Banham 1954; Zevi, ed., 1963, 1985; and Achenbach 1987 on Mendelsohn's later oeuvre. Cf. also Neutra, ed., 1986: 49. Richard Neutra reported in a letter to his wife, Dione, in October 1921: "A little while ago pictures of his new experimental neophysical building called the Einstein Tower could be seen in all the cigar stores, as well as on the cover of the *Berliner Illustrierte Zeitung*."

17. Official opinion of the Staatshochbauamt, Sept. 4, 1920; quoted in Eggers, ed., 1995: 96 (GSA Rep. 76 Uc Sec. 1, Tit. 11, part II, no. 6i, vol. 1, no. 18). For details on the planning of the many various preliminary designs and for more on the construction of the final tower, cf. Eggers, ed., 1995: 84–85; Limberg in Limberg and Staude 1994.

18. See Kirsten and Treder, eds., 1979, vol. 2, docs. 339, 351, 356, 358.

19. See Whittick 1940/56 (54): "Little could be spared beyond the demand of necessity; and the ironical fact is that with these economic restrictions Germany produced the greatest post-war architecture in the world."

20. See Forman 1973; Heilbron 1986 (86–93), Sigurdsson 1991 (82).

21. See, e.g., Hüter 1987; Banham 1960b, chap. 5, part 19: 225–29 on the "Berlin School"; 230–36 on the "Ring," founded in 1925.

22. See Pehnt 1981: 17–22, 119–25, 178.

23. Hellwag 1925: 158.

24. See Beyer, ed., 1961: 54; quoted in Eggers, ed., 1995: 79. On the many modifications of minor details of the tower, which stem from Mendelsohn's perfectionism and his close supervision of all stages of the construction, and on the many variant draft designs from the early stages of its conception, see in particular Limberg in Limberg and Staude 1994.

25. Mendelsohn's response on a questionnaire, quoted in Plaut 1929: 318.

26. See ibid. (319): "The initial sketch is right in the end. If it holds its own, this is an unmistakable and liberating sign that the work is on its way to becoming a work of art."

27. Mendelsohn as quoted in Banham 1960: 150.

28. Ibid. See also Harrington (1991), Ash (1991) on the contemporary appearance of holistic tendencies in philosophy and psychology.

29. Starting from the premise that "the layman is frightened of architecture as he is frightened of Einstein" (Robertson 1925: 72).

30. Robertson 1925: 74–75. The Einstein Tower is also called "bizarre" in Lane 1968b: 53.

31. Particularly through van Doesburg and the early *de Stijl* movement, as well as through El Lissitzky and in Russian futurism or suprematism: see Henderson 1983.

32. Its title is a clear allusion to Eddington's *Space, Time, and Gravitation* (1920). There were five editions and about 16 printings of Giedion's publication; see Chaitkin 1979: 139.

33. See Giedion 1941b: 430ff., "The new space conception: Space time," esp. pp. 436, 443.

34. See Hentschel 1990a.

35. Postscript: "Es ist einfach Klug-Scheisserei ohne jede vernünftige Basis!" Einstein to Mendelsohn, Nov. 13, 1941, KB, oversized Einstein–Mendelsohn correspondence folder.

36. Mendelsohn to Einstein, Nov. 6, 1941, KB, ibid.

37. See Chaitkin 1979: 133–34.

38. Zevi 1985: 41.

39. Mendelsohn in a lecture entitled "The International Concurrence of the Modern Building Concept; or, Dynamics and Function" ("Die Internationale Übereinstimmung des neuen Baugedankens oder Dynamik und Funktion"), in Amsterdam, 1923; reprinted in Mendelsohn 1930: 31. See also Pehnt 1981: 20.

40. Frank 1949b: 305.

41. Cf. also the detail photographs in Eggers, ed., 1995: 27–35.

42. Von Klüber to Luise Mendelsohn, enclosure to a letter of Nov. 4, 1970, KB no. I/f/39: 1–2 (original English; spelling corrected).

43. Ibid.

44. See in particular Schmidt 1930 (219–220); Banham 1960 (173: "the canonical building of expressionist architecture"), Sharp 1966 (111: "the climax of his own expressionism"), Pehnt 1981 (121–24); see also Zevi 1985 (54: "Expressionist Functionalism").

45. *Wendingen* 3, no. 10 (1920). See Staal 1920, Beyer 1920. See also, e.g., Pehnt 1981: 117, 184–85.

46. Joedicke 1966: 64.

47. Pehnt 1981: 198.

48. See, e.g., König 1966b, the fourth point on the Einstein Tower: "Expressionism. Are we really sure? . . . Well, this Swiss-watch-precise little organism could not conceivably be more remote from the nebulously social thematics of the Novembergruppe."

49. James 1994: 400. On Mendelsohn's relation to Kandinsky, see ibid. (392–94); on Freundlich's relation to Max Pechstein, see Hentschel 1995a (159–60) and above, text following n. 3.

50. See, e.g., Banham 1960b: 148, 151 n.

51. See Luise Mendelsohn to Prof. G. K. König, Aug. 16, 1966, KB no. I/f/8, point 3; and König 1966, §§4, 5.

52. According to Wolfgang Pehnt, Richard Neutra gave Steiner a guided tour of the Einstein Tower around 1922 before he built the Dornach "Goetheanum"; see Eggers, ed., 1995: 132 and n. 275. Unfortunately, despite its title, the study by the architect Karl Weidle (1929: 3) does not go into a detailed comparison of the Einstein Tower and the Goetheanum. He dismissed them both as "erudite dens" (*Gelehrtenklausen*). See, e.g., Onderdonk (1928: 238–39) for a contemporary interconnection between these two buildings.

53. As suggested by Lane 1968b: 55.

54. See Banham 1960 (147), Chaitkin 1979 (136–37), Krausse 1989 (61), James 1994 (410 n. 2).

55. See Eddy 1990; cf. with James 1994 (406–7) for the contemporary reception of futurism, and esp. of Boccioni in Paris and Berlin from 1913 onward.

56. Banham (1960: 146) mentions an "academic system" of main and secondary axes.

57. See Whittick 1940/56b (57, 193–97); Platz 1927 (70), Chaitkin 1979 (139), James 1994 (407–8).

58. Mendelsohn's 1923 lecture (see n. 39 above), according to Mendelsohn 1930 (33); cf. also James 1994 (407) for another quote from this lecture.

59. See König 1966b, point 6: "water-spots . . . fused into the wall-mass."

60. Behne 1926 (written in 1923): 38. Cf. also Peters (1926: 169) on its "stern, forward-moving forcefulness as an allegory of our times of research and industry."

61. See James (1994: 392, 398, 406) on tracing the origins of Mendelsohn's technique of depicting implied motion to Kandinsky's drawings, as well as to Italian futurism, particularly Boccioni's 1913 sculpture *Unique Forms of Continuity in Space*.

62. See Schmidt 1930: 219–20.

63. See Blossfeldt 1985.

64. Platz 1927 (70); cf. also Vischer and Hilbesheimer 1928 (17) and Frank Lloyd Wright's judgment on the Einstein Tower: "a purely plastic structure" (quoted in Onderdonk 1928: 241).

65. Pehnt 1981 (102, 121), Peters 1926 (169: "Idee des Plastischen"), Bennett 1927 (11), Hajos 1929 (497: "grandiose Betonplastik"), Schmidt 1930 (219), Banham 1960 (146).

66. Platz 1927 (70); see also Bennett 1927 (11), Onderdonk 1928 (241), and Schmidt 1930 (219) on its monolithic appearance. Cf. also Pehnt 1981 (124), and Pehnt in Eggers, ed., 1995 (132).

67. See, e.g., Staal 1920; Behne 1926 (38): "more a memorial than a laboratory."

68. Westheim 1923: 307.

69. Robertson 1925: 74–75.

70. Joël 1923: col. 2. Cf. also Peters 1926 (169) for a comparison to "tower and bridge constructions of modern capital ships."

71. Krausse 1989 (64); cf. also the illustration in Eggers, ed., 1995 (46); Krausse 1989 (117) on the periscope analogy.

72. For photos of the tower before and after the renovation, see Eggers, ed., 1995: 128–29.

73. Köster 1974: 46.

74. See König 1966 (§8), and Luise Mendelsohn's reaction to this in a letter to König, Aug. 16, 1966, KB no. I/f/8, point 4. Cf., e.g., Meyer 1928 (62–67, with pl. XV) for a contemporary description of this new trend toward functionalism, which was meant to transgress the purely representational, the rigidly dignified toward functionality, inexpensiveness, and a relaxed, uninhibited human element. See also Behne 1926 (45–47) for a critique of functionalism.

75. See, e.g., König 1966a (45), Chaitkin 1979, Krausse 1989; cf. also Neutra, ed., 1986 (54) on Neutra's involvement in building and painting the furniture.

76. See, however, the chapter on preconditions in Pehnt 1981; Lane 1968, chap. 2.

77. Mendelsohn's lecture at the Arbeitsrat für Kunst on modern architecture: "Das Problem einer neuen Baukunst," Berlin, 1919; reprinted in Mendelsohn 1930: 8. On the Arbeitsrat see Behne 1926 (9–10), Pehnt 1981 (89–106).

78. All quotes above from Mendelsohn's 1923 lecture (1930: 8–11).

79. For both versions see, e.g., Whittick 1940/56b (57), König 1966 (point 3) and the following note found among the Einstein Tower photographs at KB, folder on buildings (Bauten D-4): "Einstein Tower was completely designed in reinforced concrete. Only shortages in steel and cement forced the substitution of these materials with brick (steel and cement were being rationed). All [Mendelsohn's] English, American, and Palestinian buildings were of reinforced concrete. Some German buildings, Leningrad, and others were reinforced concrete with steel skeleton constructions."

80. See in addition Zevi 1985 (38): "There are numerous arguments in favor of brick reinforcement of a volume conceived in terms of molded and poured concrete."

81. See, e.g., Schomerus 1952 (145); see also Gössel and Leuthäuser 1990 (104), Krausse 1989 (61–62). On the history of reinforced concrete and concrete-shell construction, cf., e.g., Elliott 1994 (esp. 196–97); for contemporary, richly illustrated books on the new architectural possibilities offered by this technique see, e.g., Bennett 1927, Onderdonk 1928; and Vischer and Hilbesheimer 1928 (esp. 59–60) for the Berlin and Jena planetariums with a 6-cm-thick concrete dome cast by Dyckerhoff and Widmann in 1923.

82. Cf., Limberg and Staude 1994: 46–48, 51–54.

83. See Limberg in Limberg and Staude 1994: 51.

84. See, e.g., von Klüber 1944/48; Wempe 1975; Künzel 1986; Staude 1991, 1995; Mattig 1995: 140.

Notes to Chapter 8

1. See Hentschel 1992a.

2. On Bosch and Freundlich, see also above, Chap. 6 n. 22.

3. See, e.g., Freundlich's letter to Reichenbach, May 13, 1921, ASP no. HR-15-53-12.

4. His doctoral thesis, "Bestimmung der Dielektrizitätskonstanten von Kalium," was submitted in Marburg in 1909; see also Kirsten and Treder, eds., 1979, vol. 2, doc. 387.

5. See AAW, Astrophysical Observatory, no. 135, annual reports for 1920/21, 1921/22; the 1923 laboratory report "Arbeiten im Laboratorium des Einstein-Instituts"; and Freundlich and Hochheim 1924. See also Bottlinger 1930 on research at the Einstein Institute.

6. See Schwarzschild 1916, St. John 1917, Grebe and Bachem 1919, Bachem 1920, Kratzer 1921, Birge 1924.

7. See the letter of thanks of 1924 (undated) to BASF, addressed to the attention of the general director, Prof. Bosch, on Hochheim's "superb collaboration," AAW, Astrophysical Observatory, file no. 147.

8. Freundlich to the board of trustees of the Einstein Foundation (Kuratorium der Einstein-Stiftung), Dec. 6, 1924; reprinted in Kirsten and Treder, eds., 1979, vol. 1: 190–91. See also von der Pahlen 1926.

9. Freundlich to Reichenbach, May 13, 1921, ASP no. HR-15-53-12. On the Berlin philosopher of science Hans Reichenbach (1891–1953), see Hentschel 1990b.

10. On the solar tower in Göttingen see, e.g., ten Bruggencate and Jäger 1951; on those in Tokyo and Arcetri, see Fujita 1934 and Abetti 1926, respectively. See also Gollnow 1949 for a survey article on tower telescopes.

11. Doctorate at Göttingen University in 1911; 1920–30 lecturer at the Charlottenburg Polytechnic and on the staff of the Reich Patent Office in Berlin.

12. See, e.g., Luise Mendelsohn to Prof. F. W. Jäger, Oct. 24, 1970, KB no. I/f/30, yellow folder on buildings (Bauten D-Z), containing correspondence by Luise Mendelsohn on the Einstein Tower.

13. Franck to Einstein, Mar. 29, 1923, CPAE no. 11 268-1.

14. See Luise Mendelsohn to Prof. F. W. Jäger, Oct. 24, 1970, KB no. I/f/30.

15. Von der Pahlen was a doctoral student of Karl Schwarzschild's in Göttingen to 1909. He left Russia permanently in the aftermath of the October

Revolution, living first in Sweden, then employed as an observer at the Einstein Institute 1923–29. He is described in Freundlich 1953c as "one of the most prominent scientists among this circle," specializing in stellar statistics. In 1929 he was appointed regular professor and director of the Astronomical and Meteorological Institution at the University of Basel. On von der Pahlen see also N.N. 1952.

16. This involves the uneven distribution of radial velocities detected in the Doppler shifts of spectral class B stars. Freundlich suspected that Einstein's gravitational shift could explain why the arithmetic averages indicated various radial velocities above zero, between +4 and +5km/sec. See Freundlich and von der Pahlen 1923, von der Pahlen and Freundlich 1928.

17. Born in Potsdam, von Klüber studied at the Charlottenburg Polytechnic in Berlin and at Berlin University, where he took his doctorate under Guthnick and Miethe in 1924. He was employed at the Einstein Institute as assistant in 1923 and as observer at the Astrophysical Observatory in Potsdam from 1933. In 1941 he was appointed professor of astrophysics in Potsdam and in 1946 main observer and department head in astrophysics. In 1946 von Klüber moved away to Switzerland because of increasing difficulties arising from the Russian occupation, working for a short time at the Zurich Polytechnic, then moving to Cambridge, where he became senior observer at the university there. He observed solar magnetic fields and participated in five solar eclipse expeditions from 1929 onwards. On von Klüber see Blackwell and Dewhirst 1978.

18. After studying at Kiel, Bonn, and Munich, Brück took his doctorate in 1928 under Sommerfeld and was employed at Potsdam from 1928 to 1936. Brück moved first to Rome in 1936, then in 1937 to the Solar Physics Observatory in Cambridge, where he became John Couch Adams Astronomer and, from 1944 to 1947, also acting director. In 1947 he followed a call to Dublin at the Institute for Advanced Study and was also appointed director of the Dunsink Observatory. He later worked at the Edinburgh Royal Observatory.

19. See von Klüber 1927, Unsöld 1928, ten Bruggencate and Houtgast 1940, Gußmann 1975, Staude 1995 (144–45). See also Hentschel 1991 and references there on the Utrecht group.

20. Born in Siegen, Westphalia, Wurm studied and took his doctorate at Bonn in 1926, stayed in Potsdam until 1938, and after training at the University of Chicago's Yerkes Observatory was employed as main observer at the Hamburg Observatory in 1941. After various visits he became associate professor of astronomy at Hamburg University in 1951. On Wurm see Vanýsek 1976.

21. See Grotrian 1931, 1933; von Klüber 1931. On Grotrian, who took over the direction of the Einstein Tower in 1946 and the entire Astrophysical Observatory in Potsdam in 1950, where he had initially been employed as spectroscopist in 1922, see Freundlich 1954g, Kienle 1955, Ritschl 1955. On von Klüber see above, n. 17. Von Klüber participated in two expeditions to Sumatra to test for

light deflection in 1925/26 and 1929, as well as five other expeditions to sites of total eclipse after the war. See von Klüber 1926, 1929, 1931, 1960.

22. See Grotrian 1939; for other comments and references see also, e.g., Staude 1986, 1995 (144–45); Hufbauer 1991 (112–14); and on Grotrian's and Edlén's work, Hufbauer 1993.

23. Freundlich's progress report of the Einstein Foundation's research at the tower telescope and laboratory in the period Oct. 1, 1926, to Apr. 1, 1928: "Bericht über den Fortgang der Forschungsarbeiten der Einstein-Stiftung im Turmteleskop und Laboratorium in der Zeit vom 1. Oktober 1926 bis 1. April 1928," AAW, Astrophysical Observatory, file no. 147.

24. See St. John 1917, 1922, 1924, 1928. On St. John see also Earman and Glymour 1980a, Hentschel 1993a.

25. Freundlich's progress report on the Einstein Foundation's research, end of 1926: "Aus dem Bericht über den Fortgang der Forschungsarbeiten der Einstein-Stiftung im Turmteleskop und Laboratorium in der Zeit vom 1. April 1925 bis 1. Oktober 1926," AAW, Astrophysical Observatory, file no. 147, sheet 161.

26. For experiments on hot-cathode vacuum discharges in the Einstein Tower laboratory, see, e.g., Wolf 1927; on the electrical excitation of metal fumes in a resistance furnace, see Schüler 1926.

27. See n. 25 above.

28. Freundlich to Notgemeinschaft der Deutschen Wissenschaft, Mar. 7 (1925?), AAW, Astrophysical Observatory, file no. 151.

29. The *Yearbook of the Carnegie Institution of Washington* (25 [1925/26]: 105) mentions Freundlich's visit: "He took part in some of the solar observations on Mount Wilson and lectured on ionization and theories of radiation before the members of the staff."

30. See Freundlich's report "Bericht über den Fortgang der Forschungsar-beiten der Einstein-Stiftung im Turmteleskop und Laboratorium in der Zeit vom 1. April 1925 bis 1. Oktober 1926," AAW, Astrophysical Observatory, file no. 147: 2: "On the occasion of a visit by the undersigned to the Mount Wilson Observatory in California, where the most intense solar research has been con-ducted for 20 years, cooperation with our institute was suggested and resolved to support particularly important series of measurements at both institutes through parallel measurements at the other institute."

31. See the Mt. Wilson Observatory annual reports of the subsequent years in *Yearbook of the Carnegie Institution of Washington*, 1925/26–1932/33.

32. See BAP, Reich Interior Ministry files, no. 26769/1, sheet no. 201. Around 1934 the Notgemeinschaft der Deutschen Wissenschaft was renamed the Deutsche Forschungsgemeinschaft.

33. Freundlich to Schmidt-Ott, undated, AAW, Astrophysical Observatory, file no. 151, sheet 6.

34. Freundlich 1927b (4): "However, in commemoration of the theory of relativity, which is reinvigorating astrophysics with important questions, and in honor of its creator, Albert Einstein, it obtained the name Einstein Donation Fund, later called the Einstein Foundation."

35. Von Klüber 1965b: 82. See also von Klüber to Luise Mendelsohn, Jan. 19, 1971, KB no. I/f/31.

36. See Kirsten and Treder, eds., 1979, vol. 2, doc. 445. Scholarships at that time amounted to 150–250 marks per month, which corresponds exactly to grants of the Notgemeinschaft der Deutschen Wissenschaft in other fields to student employees. See, e.g., the list of scholarship recipients from that funding association, 1926/27, BAP, RMI, 15.01, file no. 26770/1, sheet 275; and the document cited below in n. 37. On Unsöld, see Unsöld 1972; on Johannes Picht, see also n. 38 below.

37. Freundlich to the Notgemeinschaft der Deutschen Wissenschaft, Jan. 7, 1929, AAW, Astrophysical Observatory, file no. 151 (original emphasis).

38. See Freundlich to Einstein, June 8, 1927, CPAE no. 11 292, on Picht.

39. Ibid.

40. As, e.g., in London in 1921: see Freundlich to Mendelsohn, June 13, 1921, KB.

41. See Wali's (1991: 94) biography of Chandrasekhar.

42. On this see, e.g., Schroeder-Gudehus 1978.

43. Cf., e.g., ten Bruggencate et al. 1939.

44. See Freundlich et al. 1930 for a study of the precise wavelengths of spectral lines of an iron multiplet at 72 different places on the sun's surface in comparison to its center, excluding as explanations of the center-limb variations Doppler shifts caused by convection currents and the Compton effect.

45. On Milne see, e.g., McCrea 1951, Chandrasekhar 1980.

46. See Milne to Larmor, Aug. 24, 1932, RS, Larmor Papers, Milne folder, no. 1433: "I shall be in the heart of the Einstein territory, and really want to see him and stand up to him on some matters." See also Milne 1924; Freundlich 1931d, 1933a.

47. See Crelinsten 1981 for a portrait of the American research environment in astrophysics after 1900 and Hetherington 1982 on Hubble's fight against the "eastern gang," which questioned his assertion of the direct proportionality between redshift and the distance of cosmic objects.

48. Neutra, ed., 1986: 54.

49. See the Freundlich–Einstein correspondence, CPAE nos. 11 318–11 321; as well as Freundlich to Reichenbach, Mar. 2, 1922, ASP, no. HR-15-53-04.

50. See Berliner to Einstein, Jan. 26, 1929, CPAE no. 11 301.

51. Einstein to Berliner, Jan. 30, 1929, CPAE no. 11 302.

52. Einstein to von Laue, Jan. 21, 1929, CPAE no. 11 300.

53. Einstein to Ludendorff, Sept. 15, 1925; reprinted in Kirsten and Treder,

eds., 1979, vol. 1: 196. This letter has survived only in Ludendorff's transcription, but Ludendorff's response preserved at CPAE backs the above. It adopts the same tone as the one written ten years earlier to Struve. (See above.)

Notes to Chapter 9

1. See Freundlich's (1929b, c) newspaper reports on the expedition.

2. Cf., e.g., Grotrian 1931, 1933, 1934, 1939; Hufbauer 1993; Staude in Limberg and Staude 1994: 92.

3. This was by far the largest focal length used up to that time to investigate the light deflection; see von Klüber 1960: 67.

4. See von Klüber 1932; Freundlich et al. 1931a (178), 1931b (9–10).

5. See, e.g., Freundlich et al. 1931a (180), 1931b (11–14). On the coelostat, see Freundlich et al. 1931a: 178–79.

6. The accuracy of determining stellar position shifts micrometrically from Δx_i to Δy_i could thereby be improved significantly, because micrometric errors accumulate over longer distances, only compounding any systematic errors.

7. See, e.g., Freundlich and von Brunn 1933: 222.

8. All figures from Freundlich 1930a: 314–15.

9. See in this regard Freundlich et al. 1933 as well as von Klüber 1960 (70–71), where it is described how shifts of the focal plane even as small as 1/100th of a millimeter are quite sufficient to produce such undesirable effects.

10. See in this regard Herrmann and Freundlich 1931, Meyer 1930.

11. See Freundlich et al. 1931a (173–74), 1931b (25–32); Freundlich 1955a (243–44), 1960a (124–26). For Eddington's interpretation, see Moyer 1979: 73–77.

12. Freundlich et al. 1931a: 175. See also Freundlich and von Brunn 1933 (223–29); Freundlich et al. 1933 (B14–B16).

13. See Freundlich and Gleissberg 1935; Freundlich and Ledermann 1944; Mattig 1956: 181–82.

14. Freundlich et al. 1931a (182); see also 1931b (27–30).

15. I.e., a round, dark center (the eclipsed disk) versus a bright background (the corona): see in this regard the studies by Ross 1920 at Kodak's Eastman Laboratory.

16. See Freundlich et al. 1931b: 51.

17. The displacement vectors 19 and 20, for instance, are almost perpendicular to the directly neighboring displacement vectors 26 and 30; see furthermore Freundlich et al. 1931a: 186.

18. See von Klüber 1929; Freundlich et al. 1931a, b; Freundlich 1931c; see also, e.g., Trümpler 1932, Courvoisier 1932, Danjon 1932 for differing data reductions.

19. Freundlich to Einstein, June 9, 1931, CPAE no. 11 171-1.

20. Einstein to Freundlich, June 27, 1931, CPAE no. 11 172; Freundlich to Einstein, June 29, 1931, CPAE no. 11 173.

21. See Einstein to Freundlich, July 3, 1931, CPAE no. 11 174; see also Freundlich to Einstein, July 8, 1931, CPAE no. 11 176, in which Freundlich considers Einstein's estimate too rough; and Einstein's reply to Freundlich, July 16, 1931, CPAE no. 11 177.

22. Freundlich et al. 1931a: 187–88.

23. See Freundlich et al. 1931a (190–92), 1931b (38–40, with pl. XII); von Brunn and von Klüber 1937.

24. See Dyson et al. 1920; Campbell and Trümpler 1923 / 24, 1928.

25. Freundlich et al. 1931a (194); see also 1931b (40–45).

26. Freundlich et al. 1933: B27.

27. See Ludendorff 1932a (col. 322), also in response to Jackson; Freundlich et al. 1932a, Trümpler 1932b, Danjon 1932, Jackson 1931, Freundlich and von Brunn 1933 (232–33; versus Trümpler 1932b); a (naturally biased) overview of the measurements up to and including 1929 in Dyson and Woolley 1937, chap. 7. More recent reviews include, e.g., Mattig 1956; Mikhailov 1959; von Klüber 1960; Will 1981 (167–72), 1986 (chap. 4).

28. Campbell to Trümpler, Oct. 21, 1931, copy at CPAE, no. 11 235.

29. For an overview, see Pais 1982, §17.

30. Einstein to Freundlich, June 27, 1931, CPAE no. 11 172.

31. Einstein to Freundlich, July 3, 1931, CPAE no. 11 174.

32. See, e.g., Einstein to Tolischus, Apr. 9, 1930, CPAE no. 11 163: "But consideration of the consequences of this theory is not yet far enough advanced to permit a comparison of the theory with observation."

33. See Einstein to L. Mayr, Apr. 23, 1932, CPAE no. 11 179.

34. See Ludendorff 1932a, b; Freundlich et al. 1932b.

35. See Freundlich and von Brunn 1933, Hentschel 1994.

36. Von Brunn and von Klüber 1937: 242.

37. I.e., in 1936 Mikhailov measured $E = 2''.7 \pm 0''.3$ (Mikhailov 1940); in 1947 van Biesbroeck measured $E = 2''.9 \pm 0''.3$ (van Biesbroeck 1950). See below, Table 6; see also Will 1986, chap. 4.

Notes to Chapter 10

1. Letter of the Minister of Culture to Max Wolf, June 20, 1922, Kirsten and Treder, eds., 1979, vol. 2, doc. 495: 127.

2. See Kirsten and Treder, eds., 1979, vol. 1, doc. 112; vol. 2, docs. 362, 364, 369, 370, 372, 373, 377, 380–85, 388–91, 398, 402, 413, 420, 421, 424–27, 432–34, 478–82.

3. The above details are all from the administrative settlement between the Astrophysical Observatory and the Einstein Institute: "Regelung der Geschäfts-

führung zwischen dem Astrophysikalischen Observatorium und dem Einstein-Institut," draft, signed by Medicus, dated Dec. 15, 1924, AAW, Astrophysical Observatory, file no. 149; see also final version dated Feb. 19, 1925, CPAE no. 11 280.

4. On Ludendorff, see Ludendorff 1923b, 1931; Plaut 1929 (285–86); see also Becker 1941; Münch 1941; Weber 1941; Kopff 1941 (including a bibliography of Ludendorff's publications); Brück 1942; Guthnick 1941, 1942; Strömgren 1942; Schoenberg 1948.

5. Becker 1941: 167.

6. His thesis (Ludendorff 1897) discussed the influence of small planets on Jupiter's orbit.

7. Fischer 1985: 31.

8. Quoted by Fischer (ibid.). The student employee Max Delbrück was volunteer research assistant at the tower from the winter of 1925 to the summer of 1926.

9. "As Prof. Freundlich has yet again demonstrated his ill will toward me, I must protest against his conduct." See Ludendorff's submission, Nov. 16, 1926, cited in n. 11 below. Ludendorff's resistance included refusing to support having Mendelsohn design an extension building planned in 1928, which never materialized: cf. Limberg and Staude 1994 (55–67, esp. 58 for his letter of May 19, 1928, to the ministry).

10. Becker 1941: 167.

11. See, e.g., Ludendorff's protest, dated Nov. 16, 1926, against Freundlich's submission of one of his reports "directly to the ministry, bypassing me personally" (AAW, Astrophysical Observatory, file no. 149).

12. Freundlich to Ludendorff, Jan. 7, 1925, AAW, Astrophysical Observatory, file no. 149.

13. Ludendorff to the Prussian Minister of Culture, June 29, 1931, enclosing Freundlich's application of June 29, 1931. Both documents from AAW, Astrophysical Observatory, file no. 149.

14. Ludendorff's letter dated June 29, 1931 (see above, n. 13), also shows that he was primarily concerned with finding funds for a "giant reflecting telescope for a German observatory in Spain." This project was then still very much up in the air, for reportedly Mr. Bosch "would not even consider granting the necessary enormous sums" (ibid.).

15. Ludendorff to Einstein and to the Ministry of Culture, July 2, 1928, CPAE nos. 11 281, 11 282. On Einstein's letter to Ludendorff see above, Chap. 8.

16. See Joël 1923; Ludendorff 1923/24, 1925c, 1928.

17. See Ludendorff 1932. On Kohlschütter see *Reichshandbuch der Deutschen Gesellschaft* 1 (1931): 981.

18. See AMPG, Dept. 1, Rep. 1A, no. 1659, grant dated May 17, 1923, signed by Max von Laue; approval of Undersecretary Krüss dated June 2, 1923; and Ludendorff's letter of thanks of June 4, 1923.

19. Ludendorff 1923/24: col. 77.

20. See the bibliography of Ludendorff's publications in Kopff 1941.

21. See, e.g., Ludendorff 1930.

22. Ludendorff as quoted in Plaut 1929: 285 (original emphasis). I am obliged to Mr. Bernd Ulrich for bringing this questionnaire to my attention.

23. Münch 1941: 296. For a similar assessment, see Schoenberg 1948.

24. Ludendorff as quoted in Plaut 1929: 285. Plaut published Ludendorff's confidential answers evidently without obtaining the necessary permissions.

25. Delbrück as quoted in Fischer 1985: 31.

26. Ibid.

27. Freundlich also is listed among those persons from whom Plaut had solicited responses to his questionnaire. Only a select few of the responses are quoted in the appendix, however, and Freundlich's is not among them. I have as yet been unsuccessful in attempting to locate Plaut's estate.

28. Undersecretary Richter on behalf of the Minister of Culture, ca. Mar. 17, 1931, quoted from M. von Laue's circular letter to members of the board of trustees dated Mar. 18, 1931, CPAE no. 11 309.

29. See the letter of Undersecretary Richter on behalf of the Minister of Culture to the Prussian Academy of Sciences, Mar. 20, 1933, AAW II-XIV-6, sheet 33.

30. See von Laue to Freundlich, Apr. 21, 1931, AAW, Astrophysical Observatory, file no. 148.

31. Von Laue to Freundlich, end of Apr. 1931, AAW, Astrophysical Observatory, file no. 148.

32. Von Laue to Freundlich, July 3, 1931, AAW, Astrophysical Observatory, file no. 148.

33. Freundlich to Oppenheim, Apr. 1, 1931, CPAE no. 11 312 (Oppenheim's emphasis).

34. Ibid.

35. Oppenheim to Einstein, Apr. 3, 1931, CPAE no. 11 312 verso (draft letter).

36. Freundlich to Einstein, Apr. 13, 1931, CPAE no. 11 311-1.

37. Ibid., no. 11 311-2. Reference is made in the letter to August Kopff (1882–1960); on Kopff see Gondolatsch 1961.

38. Herbert Freundlich was a member and the deputy director of the Kaiser Wilhelm Institute of Chemistry in the 1920's as well as honorary professor at both the Friedrich-Wilhelm-Universität and the Charlottenburg Polytechnic in Berlin. See Chap. 1 above.

39. See, e.g., the cover of Freundlich's personnel file, AAW, Astrophysical Observatory, file no. 152. Freundlich is entered in the 1931 edition of the Berlin biographical reference book *Reichshandbuch der Deutschen Gesellschaft* as "Dr., Prof. at the Astrophysical Observatory and head of the Einstein Institute in Potsdam." He also appears in the 1936 *List of Displaced Scholars* published in

London by the Emergency Aid Organization for German Academics in Exile (Notgemeinschaft Deutscher Wissenschaftler im Ausland): "Till 1933: Prof. Astrophysikalisches Observatorium."

Notes to Chapter 11

1. On architecture, see Lane 1968, chaps. 6–8; on astronomy, see Kuiper 1946, Kiepenheuer 1946/48; on anti-Semitic polemics against relativity theory, see Hentschel 1990a, §3.2.

2. See Mendelsohn to Luise Mendelsohn, Feb. 3, 1933, in Beyer, ed., 1961: 92.

3. This nationalist mentality motivated him to "prove" in one of his historical studies that Copernicus was a German rather than a Pole.

4. Münch 1941: 296.

5. I.e., Dr. Leist and Prof. Windelband, both in the ministry's science office.

6. Letter of Prof. Grotrian to Prof. von Laue, Apr. 7, 1933, as attachment 1 to a letter of Max von Laue to the presiding secretary of the Prussian Academy of Sciences and to the advisory board of the Potsdam Astrophysical Observatory, located at AAW, no. II–XIV-6, sheet 34.

7. Ibid.

8. Transcription of a letter of Ludendorff to the Minister of Culture, Apr. 6, 1933, AAW, no. II–XIV-6, sheet 36 (= attachment 3 to von Laue's letter cited above in n. 6).

9. Vahlen to Rust, Mar. 28, 1933; quoted from Kirsten and Treder, eds., 1979, vol. 1, doc. 115: 196.

10. Vahlen was first district leader of the Pomeranian Nazi Party from 1924 and was suspended from his academic appointment in 1927 for ordering the removal of the black, red, and gold national flag from the Greifswald University building. On Vahlen see Siegmund-Schultze 1984; Walker 1995, chaps. 4, 5.

11. An example of Freundlich's countermeasures is his petition to the Minister of Culture of Oct. 2, 1933(!), to fill a vacated associate position at the observatory with one of his coworkers, Prof. Albert von Brunn: AAW, Astrophysical Observatory, file no. 149. Investigations at the tower were turned toward studying solar granulation, and in the autumn of 1941 a regular series of observations of the sun's magnetic field was started, especially on the Zeeman effect in solar spots; see von Klüber 1944/48; Grotrian 1952: 89–92. On Albert von Brunn, see Freundlich 1951b.

12. Freundlich's note at the bottom of Ludendorff's circular of Oct. 5, 1933, AAW, Astrophysical Observatory, file no. 149. Cf. the facsimile in Hentschel 1992c: 149.

13. Ludendorff to Freundlich, Oct. 7, 1933, AAW, no. II–XIV-6, sheet 45.

14. For a facsimile of the article (Freundlich 1934b), see Hentschel 1992c: 150.

15. See Ludendorff to Freundlich (in Istanbul), June 23, 1934, AAW, Astrophysical Observatory, file no. 152. On the above law and its impact on the German scientific community see Hentschel, ed., 1996, doc. 7, and §3 of the introduction there.

16. Freundlich to the Potsdam Astrophysical Observatory, July 3, 1934, AAW, Astrophysical Observatory, file no. 152.

17. On the emigration of scientists in general, see, e.g., Bentwich 1953, Hoch 1983.

18. The circumstances under which German scholars, politicians, and artists found asylum on the Bosporus are described by Neumark 1980 and Widmann 1973.

19. See the transcription of Freundlich's letter to the Minister of Culture, Oct. 3, 1933, AAW, II-XIV-6, sheet 46, attachment 2.

20. Ibid.

21. See the transcription of Freundlich's letter to the Minister of Culture, Feb. 6, 1934, AAW, no. II-XIV-6, sheet 48, attachment 4.

22. "He is not to be notified of this. Should he put forward a claim there, then I am to be informed first": Vahlen on behalf of the Minister of Culture, Apr. 3, 1934, AAW, Astrophysical Observatory, file no. 152. See also the personnel file on Freundlich in HUBU regarding his fight for payment of his salary and pension up to 1934. The retirement date recorded there is July 1, 1934.

23. Freundlich (in Istanbul) to the Prussian Academy of Sciences, Feb. 6, 1934, AAW, no. II-XIV-6, sheet 44.

24. Strohbusch was employed at the Einstein Tower from February 1925 until well into the 1960's and was succeeded by Herbert Borchert, who also worked there for about 50 years.

25. P. ten Bruggencate took his doctorate under von Seeliger in Munich, was teaching assistant to Kienle in Göttingen for a short time, spent time at various foreign observatories and institutes between 1926 and 1929, and became university lecturer in Greifswald in 1929. On ten Bruggencate, who became a member of the Nazi Party in 1937 and was a member of the SA 1933–41, see, e.g., Unsöld 1961 (60–61), Jäger 1961, Kienle 1962, Litten 1992 (256).

26. See Wolfschmidt 1994.

27. On von Klüber, see above, Chap. 8 n. 17.

28. On this magnetic field research, which was continued after the war (from 1947 onwards under the direction of Walter Grotrian) see, e.g., Grotrian 1952 (89–90), Künzel 1986, Staude 1991a (506–7). On the Einstein Tower in the Third Reich, see the quote in text below at n. 39.

29. See Gleissberg 1967 (275), a report on the condition of astronomy in Turkey.

30. Gleissberg 1967 (277); see also Widmann 1973 (95), and on the Zeiss astrograph Schomerus 1952 (139).

31. See Freundlich 1934/35, 1935b–e, 1936, 1937b.

32. Freundlich's (1937b) astronomy textbook was translated into Turkish with the help of his assistant, Gleissberg, and the future physics professor Yenicay.

33. Freundlich had been proposed by Pollak; see Pollak to Einstein, Mar. 20, 1935, CPAE no. 11 180; and Einstein's personal reference, Apr. 10, 1935, CPAE no. 11 182. See also Hoffmann 1991 and Freundlich's (1937a) only paper to appear in Czech. His successor in Istanbul was Hans Rosenberg from Kiel University, who died shortly afterwards, however, in 1940 from heatstroke; see Neumark 1980 (100), Widmann 1973 (95).

34. See Batten 1985: 33. Batten assesses Freundlich's teaching abilities as follows: "While he has achievements to his credit in other fields of research, I believe that his influence on his students was one of his most important contributions to astronomy, and I have tried to emphasize how that spread to many countries" (ibid.: 35). See also Freundlich's (1953d) own article on the didactic value of astronomy.

35. See Forbes 1972: 183.

36. See the progress reports during his stay there (Freundlich 1943–59) and his letters to Mendelsohn dated 1946 and later at KB. See also N.N. 1959.

37. See Donnan 1942/44: 30, 32.

38. See Freundlich 1945b.

39. Freundlich to Mendelsohn, Aug. 12, 1946, KB (original English, apart from the Kellermann quote, which appears here in translation).

40. See Freundlich 1950b, Jack 1951.

41. See Batten (1985: 34) for more details. See also Freundlich 1953d for his inaugural lecture as first Napier Professor of Astronomy at the University of St. Andrews.

42. See Batten 1985: 34.

43. A mechanic Freundlich had hired, Robert L. Waland, who according to Batten (1985: 34) had originally been an amateur astronomer whom Freundlich had succeeded in convincing to become an independent instrument maker, resigned as a direct result of these tensions before installation of the optics was complete. This explains in part why the 36-inch telescope was inoperational for so long and why it never actually did achieve its full potential in producing quality images. See Forbes 1972: 184.

Notes to Chapter 12

1. See, e.g., Freundlich 1953a, 1954a–f, 1957; Freundlich and Forbes 1956, 1959.

2. See in this regard, e.g., the comments by Born 1954, McCrea 1954, and ter Haar 1954, as well as Hopmann 1954, Ferrari d'Occhieppo 1954, and Melvin 1955.

3. See, e.g., Burbidge 1954; and Batten 1985 (35) for a description of a

heated debate between Born and Freundlich on the occasion of Born's farewell lecture in St. Andrews, which culminated in a "fever-pitch of his excitement" and Freundlich's subsequent heart attack. Born wrote laconically to Einstein on Jan. 20, 1954: "Freundlich was, by the way, very ill—coronary thrombosis"; see Born, ed., 1969: 286. Cf. also Hentschel 1995b, chap. 10, for a more detailed report on the fate of Freundlich's theory.

4. See also Born to Einstein, May 4, 1952, Born, ed., 1969 (255–56): "Yesterday Freundlich was here and held a clear lecture on the state of light deflection by the sun. It really does look as if your formula isn't quite right. It looks even worse for redshift: within the sun's disk it is much smaller than the theoretical value; and at the rim, larger. What's going on here? Could this be an indication of nonlinearity (scattering of light by light)? Have you considered this?"

5. Einstein to Born, May 12, 1952, Born, ed., 1969: 258.

6. See RAS Papers 81/2, Minutes of the Joint Eclipse Committee, Nov. 14, 1952, July 2, 1953, Aug. 27, 1954, and Jan. 14, 1955. Freundlich received contributions from the Royal Astronomical Society of £600 and £500. See also Freundlich 1954g (441) on Grotrian's planned participation; von Klüber 1955.

7. CPAE no. 11 195.

8. Batten 1985: 35.

9. See Freundlich 1931/32, 1951a; see also Hentschel 1990a (§§3.4, 4.7), 1990b.

10. On the history of solar redshift between 1890 and 1960, see Hentschel 1995.

11. For modern high-precision tests of general relativity theory see, e.g., Will 1986 and references there.

12. See also Eisenstaedt 1989.

Notes to Chapter 13

1. On the crisis of German science caused by the inflation see, e.g., *Internationale Monatsschrift für Wissenschaft, Kunst und Technik* 15 (1920), nos. 1, 2.

2. See in this regard Hentschel 1990a, §3.4.

3. See, e.g., Freundlich to Mendelsohn, Aug. 12, 1946, KB, quoted above in Chap. 11, text at n. 39.

4. Freundlich to Mendelsohn, Feb. 25, 1946, KB (original English).

5. Cf. in particular Galison 1993 on the broad cultural meaning of *Aufbau* in many different cultural contexts in the interwar period, ranging from philosophy to architecture.

6. On Einstein's reception see, e.g., Glick, ed., 1987; Hentschel 1990. On Mendelsohn's reception in terms of "modernity" see, e.g., Peters 1926 (169): "because all memories of past events have sunk into oblivion, something new to the core [emerged,] the highly personal achievement of a modern person who

has learned to seek from the start a new design for new functions." For a more critical view of modern in contrast to traditional architecture, see Meyer 1928: 8.

7. The Astrophysical Observatory was founded in 1879, in accordance with the suggestion of W. Foerster in 1871. See also Foerster 1911, chap. 17; Herrmann 1975.

8. For an overview of Berlin scientific institutions, see, e.g., Buddensieg et al., eds., 1987, vol. *Disziplinen*, pp. 44–49; vol. *Objekte*, pp. 262–63. The most commendable overview on the Einstein Tower's foundation and operation is still Freundlich 1927b; see also Freundlich 1925b.

9. Herrmann 1973 is an interesting comparison between the United States and Germany; see also Herrmann and Hamel 1975.

10. For a survey of industry-backed research support in Germany, see Richter 1979.

11. See Hughes 1986.

12. Cf. Price 1963 and Trischler 1992 (280–81) on a parallel in aerodynamics research, where within the eight years between 1930 and 1938 instrument expenses rose by a factor of 10.

13. On the spread of scientific internationalism among Weimar physicists, see in particular Forman 1973.

14. See, e.g., Freundlich 1919a, 1920a, 1928d, 1929b, popular articles for Berlin newspapers such as *Vossische Zeitung*, *Berliner Tageblatt*, or *Berliner Illustrierte Zeitung*; but see also Kirchberger 1925 in the widely read *Frankfurter Zeitung* and Fürst 1926 for *Koralle: Magazin für alle Freunde von Natur und Technik*.

15. The conflicts on the foundations of mathematics have striking similarities, as described by Mehrtens 1990: Einstein's role is in many ways analogous to Hilbert's role as *Generaldirektor*, wherein he gathered a mass of people around himself to carry out his program, while his opponents made an appeal for intuition.

16. See N.N. 1926.

17. On these crossover cases within the system of mathematics, see again Mehrtens 1990 (394–95); on "reactionary modernism" in the context of technology, literature, philosophy, and politics, see in particular Herf 1984.

18. See Stone's (1990) photographic works from 1925 to 1939.

19. Von Klüber in an enclosure to a letter to Luise Mendelsohn, Nov. 4, 1970 (original English), KB no. I/f/39, p. 2.

20. Hellwag 1925 (157); cf. in a similar tone Schmidt 1930 (220): an "audacious gamble."

21. Mendelsohn 1930: 34.

Abbreviations in the References

Abh.Berlin	*Abhandlungen der Preußischen Akademie der Wissenschaften*, Berlin
AdA	*Annales d'Astrophysique*
AIHS	*Archives Internationales d'Histoire des Sciences*
AN	*Astronomische Nachrichten*
Ann.Phys.	*Annalen der Physik*
ApJ.	*Astrophysical Journal*
BEKSB	*Beobachtungs-Ergebnisse der Königlichen Sternwarte Berlin*
CRAS	*Comptes Rendus Hebdomadaires des Séances de l'Académie des Sciences*
DF	*Deutsche Forschung*
DSB	*Dictionary of Scientific Biography*
EEN	*Ergebnisse der Exakten Naturwissenschaften*
FuF	*Forschungen und Fortschritte: Korrespondenzblatt Deutscher Wissenschaft und Technik*
HdA	*Handbuch der Astrophysik*
HdP	*Handbuch der Physik*
HSPS	*Historical Studies in the Physical Sciences*
JRE	*Jahrbuch der Radioaktivität und Elektronik*
MAG	*Mitteilungen der Astronomischen Gesellschaft*
MNRAS	*Monthly Notices of the Royal Astronomical Society*
Natw.	*Die Naturwissenschaften*
NTM	*Schriftenreihe für Geschichte der Naturwissenschaften, Technik und Medizin*
NTNI	*Natuurkundig Tijdschrift voor Nederlandsch-Indie*
NYT	*New York Times*
Obs.	*Observatory*
ONFRS	*Obituary Notices of Fellows of the Royal Society*
PB	*Physikalische Blätter*
Phil.Mag.	*London, Edinburgh and Dublin Philosophical Magazine*

Phys. Rev.	*Physical Review*
Phys. Z.	*Physikalische Zeitschrift*
PIUO	*Publications of Istanbul University, Observatory*
PNAS	*Proceedings of the National Academy of Sciences*
Pop. Astr.	*Popular Astronomy*
PPSL	*Proceedings of the Physical Society*, London
QJRAS	*Quarterly Journal of the Royal Astronomical Society*
SB. Berlin	*Sitzungsberichte der Preußischen Akademie der Wissenschaften*, Berlin, Math.-phys. Klasse
SB. München	*Sitzungsberichte der Bayerischen Akademie der Wissenschaften*, Munich, Math.-phys. Klasse
SDM	*Süddeutsche Monatshefte*
VAG	*Vierteljahresschrift der Astronomischen Gesellschaft*
ViA	*Vistas in Astronomy*
VNGZ	*Vierteljahresschrift der Naturforschenden Gesellschaft*, Zurich
VZ	*Vossische Zeitung*
Wiss. Ann.	*Wissenschaftliche Annalen*
ZAP	*Zeitschrift für Astrophysik*
ZfI	*Zeitschrift für Instrumentenkunde*
Z. Phys.	*Zeitschrift für Physik*

References

Abetti, Giorgio. 1926. La torre solare. *Pubblicazioni della R. Università degli Studi di Firenze* no. 43: 11–22; also *Ingegneria* 5, no. 1: 2–7.

———. 1929. Solar physics. *HdA* 4: 57–230.

Abraham, Max. 1912. Zur Theorie der Gravitation. *Phys.Z.* 13: 1–14.

———. 1914. Neuere Gravitationstheorien. *JRE* 11: 470–520.

Achenbach, Sigrid. 1987. *Erich Mendelsohn, 1887–1953: Ideen, Bauten, Projekte.* Berlin: Staatliche Museen Preußischer Kulturbesitz. (Catalog of the centenary exhibition from the art museum's holdings.)

Adams, Walter S. 1938. George Ellery Hale. *APJ.* 87: 369–88.

Ash, Mitchell. 1991. Gestalt psychology in Weimar culture. *History of Human Sciences* 4: 395–415.

Auwers, Arthur. 1905. Wahlvorschlag für Hugo Ritter von Seeliger. In Kirsten and Körber, eds., 1975, 1: 169–71.

Bachem, Albert. 1920. Magnetoptische Untersuchungen an der Stickstoffbande 3883. *Z.Phys.* 3: 372–88.

Banham, Reyner. 1954. Mendelsohn. *The Architectural Review* 116, no. 692 (Aug.): 84–93.

———. 1960. *Theory and design in the first machine age.* (a) New York: Praeger (see esp. pp. 173–75 on the Einstein Tower); (b) in German translation *Die Revolution der Architektur* Hamburg: Reinbek, 1964).

Batten, Alan H. 1985. Erwin Finlay-Freundlich, 1885–1964. *Journal of the British Astronomical Association* 96: 33–35.

Bauersfeld, Walther. 1925. *Das Zeissische Projektions-Planetarium.* Jena: Jozewski.

Becker, Wilhelm. 1941. "Professor Dr. Hans Ludendorff †." *Sterne* 21: 166–67.

Behne, Adolf. 1926. *Der moderne Zweckbau.* Munich: Drei-Masken-Verlag. Reprinted Frankfurt: Ullstein, 1964. (Written in 1923; see esp. pp. 37, 45.)

Bennett, Thomas Penberthy. 1927. *Architectural design in concrete.* London: Benn. (See esp. pp. 11 and lxii–lxiii.)

Bentwich, Norman. 1953. *The rescue and achievement of refugee scholars: The story of displaced scholars and scientists, 1933–1952.* The Hague: Nijhoff. (See esp. pp. 53–56 on emigration to Turkey.)

Beyer, Oskar. 1920. Architectuur in Ijzer en Beton. *Wendingen* 3, no. 10: 4–14. (See also Staal 1920.)

———, ed. 1961. *Erich Mendelsohn: Briefe eines Architekten.* Munich: Prestel.

van Biesbroeck, G. 1950. The Einstein shift at the eclipse of May 20, 1947, in Brazil. *Astronomical Journal* 55: 49–54. (See also reply in Freundlich 1950a.)

———. 1953. The relativity shift at the 1952 eclipse of the sun. *Astronomical Journal* 58: 87–88.

Birge, Raymond Thayer. 1924. The 3883 cyanogen band in the solar spectrum. *ApJ.* 59: 45–60.

Blackwell, D. E., and D. W. Dewhirst. 1978. H. von Klüber. *Nature* 273: 414.

Blossfeldt, Karl. 1985. *Art forms in nature.* London: A. Zwenner, 1967. Reprinted New York: Dover, 1985.

Blumenthal, Otto. 1918. Karl Schwarzschild. *Jahresberichte der Deutschen Mathematiker-Vereinigung* 26: 56–75.

Born, Max. 1954. On the interpretation of Freundlich's redshift formula. *PPSL* A67: 193–94. (Reply to Freundlich 1954b: 192–93.)

———, ed. 1969. *Albert Einstein—Hedwig und Max Born Briefwechsel, 1916–1955.* Munich: Nymphenburger.

Bottlinger, K. F. 1930. Zehn Jahre Einstein-Institut. *Natw.* 18: 777–78.

Briggs, John W. 1991. Rebirth of the Snow telescope. *Sky and Telescope* 81: 206–11.

Brück, Hermann Alexander. 1936. Spektralphotometrie. *HdA* 7: 51–70.

———. 1942. Friedrich Wilhelm Hans Ludendorff. *MNRAS* 102: 78–79.

ten Bruggencate, Paul. 1930. The radial velocities of globular clusters. *PNAS* 16: 111–18. (See also comment in von der Pahlen and Freundlich 1930.)

———. 1955. Karl Schwarzschild. In Hans Schwerte and Wilhelm Spengler, eds. *Forscher und Wissenschaftler im heutigen Europa:* 232–39. Oldenburg: Stalling.

ten Bruggencate, P., and Jakob Houtgast. 1940. Bestimmung der Dämpfung zweier Fe-Multiplette für Mitte und Rand der Sonnenscheibe. *ZAP* 20: 149–71. (MAOP no. 7.)

ten Bruggencate, P., et al. 1939. [P. ten Bruggencate, J. Houtgast, and H. von Klüber.] Beobachtungs- und Reduktions-Technik bei den Potsdamer spektralphotometrischen Untersuchungen von Fraunhoferlinien. *Publikationen des Astrophysikalischen Observatoriums, Potsdam,* 29, no. 3: 1–21.

ten Bruggencate, P., and F. W. Jäger. 1951. Das Turmteleskop der Göttinger Sternwarte. *Nachrichten der Akademie der Wissenschaften in Göttingen,* Math.-phys. Klasse, 5: 1–28.

ten Bruggencate, P., and H. H. Voyt. 1958. Das Turmteleskop der Göttinger Sternwarte, II (Instrumentelle Ergänzungen und Verbesserungen). Ibid. 12: 125–211.

von Brunn, Albert. 1930. Zur Frage der 'Rotverschiebung.' *Sterne* 10: 166–71. (See also Freundlich et al. 1930.)

von Brunn, A., and H. von Klüber. 1937. Kritische Untersuchung zur Bestimmung der Lichtablenkung durch die Potsdamer Sonnenfinsternisexpedition von 1929. *ZAP* 14: 242–50.

Buddensieg, Tillman, et al., eds., 1987. [T. Buddensieg, K. Düwell, and K.-J. Sembach.] *Wissenschaft in Berlin.* 3 vols. Berlin: Mann.

Burbidge, Eleanor Margaret, and Geoffrey R. Burbidge. 1954. On the observational data relating to Freundlich's proposed red-shift law. *Phil.Mag.,* 7th ser., 45: 1019–22. (Reply to Freundlich 1954a; see also ter Haar 1954, Kirschstein 1954.)

Campbell, William Wallace. 1911a. On the motions of the brighter class B stars. *Lick Observatory Bulletin* 6: 101–24.

———. 1911b. Some peculiarities in the motion of stars. Ibid. 6: 125–35.

———. 1914. Note on evidences of rapid convection in stellar atmospheres. *Lick Observatory Bulletin* 8: 82.

Campbell, W. W., and R. J. Trumpler. 1923/24. Observations on the deflection of light in passing through the sun's gravitational field. *Lick Observatory Bulletin* 11: 41–54.

———. 1928. Observations made with a pair of five-foot cameras on the light deflections in the sun's gravitational field at the total solar eclipse of Sept. 21, 1922. *Lick Observatory Bulletin* 13: 130–60.

Castagnetti, Giuseppe, et al., eds., 1994. [G. Castagnetti, P. Damerow, W. Heinrich, J. Renn, and T. Sauer.] Wissenschaft zwischen Grundlagenkrise und Politik: Einstein in Berlin. Berlin: MPI für Bildungsforschung. (Working report of the Arbeitsstelle Albert Einstein 1991–1993.)

Chaitkin, Bill. 1979. Einstein and architecture. In Maurice Goldsmith, ed., *Einstein: The first hundred years:* 131–44. Oxford: Pergamon Press.

Chandrasekhar, Subrahmanyan. 1980. Edward Arthur Milne: His part in the development of modern astrophysics. *QJRAS* 21: 93–107.

Chant, C. A., and R. K. Young. 1924. Evidence of the bending of the rays of light on passing the sun, obtained by the Canadian expedition to observe the Australian eclipse. *Publications of the Dominion Astrophysical Observatory* 2, no. 15: 275–85 and plates.

Clark, Ronald W. 1971. *Einstein: The life and times.* New York: World Publishing Company. Reprinted London: Hodder and Stoughton, 1973.

Courvoisier, Leo. 1905. Kinemaras Phänomen und die 'jährliche Refraktion' der Fixsterne. *AN* 167: cols. 81–106.

———. 1913. Über systematische Abweichungen der Sternpositionen im Sinne einer jährlichen Refraktion. *BEKSB* 15: 1–79. (See also Freundlich 1913d.)

———. 1919. Neue Untersuchungen über die jährliche Refraktion. *AN* 209: cols. 337–56.

———. 1920a. Jährliche Refraktion und Sonnenfinsternisaufnahmen 1919. *AN* 211: cols. 305–12.

———. 1920b. Hermann Struve. *AN* 212: cols. 33–38.

———. 1932. Sonnenfinsternis-Aufnahmen 1919, 1922, 1929 und jährliche Refraktion. *AN* 244: cols. 275–88. (Comment on Freundlich et al. 1932b.)

Crelinsten, Jeffrey Michael. 1981. The reception of Einstein's general theory of relativity among American astronomers, 1910–1930. Ph.D. thesis, University of Montreal.

———. 1983. William Wallace Campbell and the "Einstein problem": An observational astronomer confronts the theory of relativity. *HSPS* 14: 1–92.

Cunningham, Ebenezer. 1920. Review of Freundlich 1916c. *Nature* 105: 350–51.

Danjon, André. 1932. Le déplacement apparent des étoiles autour du soleil éclipsé. *Journal de Physique*, 7th ser., 3: 281–301. (Abstract in *CRAS* 194: 252–54.)

Diecke, Sally H. 1975. Schwarzschild, Karl. *DSB* 12: 247–53.

Dodwell, G. F., and C. R. Davidson. 1924. Determination of the deflection of light by the sun's gravitational field from observations made at Cordillo Downs, South Australia, during the total eclipse of 1922. *MNRAS* 84: 150–62.

Donnan, F. G. 1942/44. Freundlich, Herbert. *ONFRS* 4: 26–50.

Dyson, Frank Watson. 1921. Karl Hermann Struve. *MNRAS* 81: 270–72.

———. 1939. William Wallace Campbell, 1862–1938. *ONFRS* 2: 613–19.

Dyson, F. W., et al. 1920 [F. W. Dyson, A. S. Eddington, and C. Davidson.] A determination of the deflection of light of the sun's gravitational field from observations made at the total eclipse of May 29, 1919. *Philosophical Transactions of the Royal Society, London*, ser. A, 220: 291–334. (See also Freundlich 1920e.)

Dyson, F. W., and Richard van der Riet Woolley. 1937. *Eclipses of the sun and moon*. Oxford: Clarendon Press. (See esp. chap. 7 on light deflection.)

Earman, John, and Clark Glymour. 1980a. The gravitational redshift as a test of general relativity: History and analysis. *Studies in History and Philosophy of Science* 11: 251–78.

———. 1980b. Relativity and eclipses: The British eclipse expeditions of 1919 and their predecessors. *HSPS* 11: 49–85.

von Eckhardt, Wolf. 1960. *Eric Mendelsohn*. New York: Braziller. (See esp. p. 14 on the Einstein Tower.)

Eddington, Arthur Stanley. 1915. Some problems of astronomy, 19: Gravitation. *Obs.* 38: 93–98. (See also comment in Freundlich 1916f.)

———. 1920. *Space, time, and gravitation: An outline of the general relativity theory*. Cambridge: Cambridge University Press. (See also review in Harrow 1921.)

———. 1923. *The mathematical theory of relativity*. Cambridge: Cambridge University Press.

Eddy, David Hamilton. 1990. Perspective: The secret futurist [Erich Mendelsohn.] *Times Higher Education Supplement*, Aug. 24: 11, 13.

Eggers, Barbara, ed. 1995. *Der Einsteinsturm in Potsdam: Architektur und Astrophysik.* Berlin: ARS Nicolai. (Published by the Astrophysikalisches Institut Potsdam.)

Eggert, W. 1914. Der Neubau der Berliner Sternwarte auf dem Babelsberg. *Zeitschrift für Bauwesen* 64: cols. 645–74 and pls. 54–59.

Einstein, Albert. 1905. Zur Elektrodynamik bewegter Körper. (a) *Ann.Phys.*, 4th ser., 17: 891–921; (b) in Sommerfeld, ed., 1923: 26–50; (c) in Stachel et al., eds., 1989, doc. 23: 275–310. (English trans. by A. Beck, ibid. pp. 140–71.)

——. 1906. Ist die Trägheit eines Körpers von seinem Energieinhalt abhängig? (a) *Ann.Phys.*, 4th ser., 18: 639–41; (b) in Sommerfeld, ed., 1923: 51–53; (c) in Stachel et al., eds., 1989, doc. 24: 311–15 (English trans. by A. Beck, ibid. pp. 172–74).

——. 1906d. Das Prinzip von der Erhaltung der Schwerpunktsbewegung und die Trägheit der Energie. *Ann.Phys.*, 4th ser., 20: 627–33. (e) In Stachel et al., eds., 1989, doc. 35: 359–66. (English trans. by A. Beck, ibid. pp. 200–206.)

——. 1907a. Über die vom Relativitätsprinzip geforderte Trägheit der Energie. *Ann.Phys.*, 4th ser., 23: 371–84. (b) In Stachel et al., eds., 1989, doc. 45: 413–28. (English trans. by A. Beck, ibid. pp. 238–50.)

——. 1907c. Über das Relativitätsprinzip und die aus demselben gezogenen Folgerungen. *JRE* 4: 411–62. (d) In Stachel et al., eds., 1989, doc. 47: 432–88. (English trans. by A. Beck, ibid. pp. 252–311; see also Einstein 1908.)

——. 1908. Correction of Einstein 1907c. (a) *JRE* 5: 98–99; (b) in Stachel et al., eds., 1989, doc. 49: 493–95 (English trans. by A. Beck, ibid. pp. 316–17).

——. 1911. Über den Einfluß der Schwerkraft auf die Ausbreitung des Lichtes. (a) *Ann.Phys.*, 4th ser., 35: 898–908; (b) in Sommerfeld, ed., 1923: 72–80; (c) in Klein et al., eds., 1993a, doc. 23: 485–97 (English trans. by A. Beck, ibid. pp. 379–87).

——. 1912a. Lichtgeschwindigkeit und Statik des Gravitationsfeldes. *Ann.Phys.*, 4th ser., 38: 355–69.

——. 1912b. Zur Theorie des statischen Gravitationsfeldes. Ibid., 4th ser., 38: 443–58.

——. 1912c. Relativität und Gravitation: Erwiderung auf eine Bemerkung von M. Abraham. Ibid., 4th ser., 38: 1059–64. (Reply to Abraham 1912.)

——. 1912d. Bemerkung zu Abrahams Auseinandersetzung: Nochmals Relativität und Gravitation. Ibid., 4th ser., 39: 704.

——. 1912e. Die Relativitätstheorie. *VNGZ* 56: 1–14.

——. 1913a. Physikalische Grundlagen einer Gravitationstheorie. *VNGZ* 58: 284–90.

——. 1913b. Zum gegenwärtigen Stande des Gravitationsproblems. *Phys.Z.* 14: 1249–66; and ibid. 15: 108.

——. 1914a. Prinzipielles zur verallgemeinerten Relativitätstheorie und Gravitationstheorie. Ibid. 15: 176–80.

——. 1914b. Zum Relativitätsproblem. *Scientia* 15: 332–48. (Reprinted ibid. 112 [1977]: 1–16.)

——. 1914c. Zur Theorie der Gravitation. *VNGZ* 59: iv–vi.

——. 1914d. Formale Grundlagen der Allgemeinen Relativitätstheorie. *SB.Berlin*: 1030–85.

——. 1915a. Zur allgemeinen Relativitätstheorie. Ibid.: 778–86; with Nachtrag ibid.: 799–801.

——. 1915b. Erklärung der Perihelbewegung des Merkur aus der allgemeinen Relativitätstheorie. Ibid.: 831–39.

——. 1915c. Die Feldgleichungen der Gravitation. Ibid.: 844–47.

——. 1915/25. Die Relativitätstheorie. In *Kultur der Gegenwart: Die Physik.* Leipzig and Berlin: Teubner. (**a**) 1st ed. (1915): pp. 703–13; (**b**) 2d ed. (1925): pp. 783–97.

——. 1916. Die Grundlagen der allgemeinen Relativitätstheorie. (**a**) *Ann.Phys.*, 4th ser., 49: 769–822; (**b**) offprint (Leipzig: Barth); (**c**) in Sommerfeld, ed., 1923: 81–124.

——. 1916d. Gedächtnisrede auf Karl Schwarzschild. *SB.Berlin*: 768–70.

——. 1917. *Über die spezielle und die allgemeine Relativitätstheorie.* Braunschweig: Vieweg. (**a**) 1st ed. (1917: Tagesfragen aus den Gebieten der Naturwissenschaften und der Technik, no. 38); (**b**) 10th ed. (1920); (**c**) reprint of (1969) 21st ed. (1979); (**d**) English trans. as *Relativity: The special and general theory* (New York: Crown, 1961).

——. 1917e. Kosmologische Betrachtungen zur Allgemeinen Relativitätstheorie. *SB.Berlin*: 142–52. (Cf. *Natw.* 7: 232.) (**f**) In Sommerfeld, ed., 1923: 130–39.

——. 1920. *Relativity.* New York: Holt. (See also review in Harrow 1921.)

Einstein, A., and A. D. Fokker. 1914. Die Nordströmsche Gravitationstheorie vom Standpunkt des absoluten Differentialkalküls. *Ann.Phys.*, 4th ser., 44: 321–28.

Einstein, A., and Marcel Grossmann. 1913. Entwurf einer verallgemeinerten Relativitätstheorie und einer Theorie der Gravitation. (**a**) *Zeitschrift für Mathematik und Physik* 62: 25–61; (**b**) offprint (Leipzig: Teubner). (Part 1, physics, A. Einstein; Part 2, mathematics, M. Grossmann.)

Eisenstaedt, Jean. 1989. The low-water mark of general relativity, 1925–1955. In Howard and Stachel, eds., 1989: 277–92.

Elliott, Cecil D. 1994. *Technics and architecture: The development of materials and systems for buildings.* Cambridge, Mass.: M.I.T. Press.

Eulenburg, Franz. 1920. Die Preisrevolution seit dem Krieg. *Jahrbücher für Nationalökonomie* 115 (3d ser., vol. 60): 289–338.

Evershed, John. 1913. A new interpretation of the general displacements of the lines of the solar spectrum towards the red. *Kodaikanal Bulletin* 3, no. 36: 45–53.

Ewald, Peter Paul. 1960. Max von Laue. *Biographical Memoirs of Fellows of the Royal Society, London*, 6: 135–56.

Fabry, Charles, and Henri Buisson. 1910. Application de la méthode interféren-tielle à la mesure de très petits déplacements de raies: comparaison du spectre solaire avec le spectre d'arc du fer—comparaison du centre et du bord du soleil. *AP.J.* 31: 97–119.

Ferrari d'Occhieppo, K. 1954. A possible test for Freundlich's red-shift. *Obs.* 74: 169–70.

Finlay-Freundlich, Erwin. *See* Freundlich, Erwin F.

Fischer, Peter. 1985. *Licht und Leben: Ein Bericht über Max Delbrück, den Wegbereiter der Molekularbiologie*. Constance: Universitätsverlag.

Foerster, Wilhelm. 1911. *Lebenserinnerungen und Lebenshoffnungen*. Berlin: Reimer.

Forbes, Eric Gray. 1963. A history of the solar red shift problem. *Annals of Science* 17: 129–64.

——. 1972. Freundlich, Erwin Finlay. *DSB* 5: 181–84.

Forman, Paul. 1973. Scientific internationalism and the Weimar physicists. *Isis* 64: 151–80.

——. 1974. The financial support and political alignment of physicists in Weimar Germany. *Minerva* 12: 39–66.

Franck, James. 1960. Max von Laue. *Yearbook of the American Philosophical Society*: 155–59.

Frank, Philipp. 1949. *Einstein: Sein Leben und seine Zeit.* (**a**) 1st ed. (Munich: List, 1949); (**b**) 2d ed. (Braunschweig: Vieweg, 1979); (**c**) transl. from the German manuscript by George Rosen, as *Einstein: His life and times*, ed. Shuichi Kusaka (Norwalk, Conn.: Easton Press, 1988); (**d**) paperback (New York: Da Capo Press, 1989).

Freundlich, Erwin F. [Later: Erwin Finlay-Freundlich] 1911. Analytische Funk-tionen mit beliebig vorgeschriebenem unendlich vielblättrigem Existenz-bereiche. Thesis, University of Göttingen. (Completed Jan. 1910 under P. Koebe.)

——. 1913a. Die Lichtkurve der Nova Geminorum 2 nach den Beobachtungen auf der königlichen Sternwarte Berlin. *AN* 193: cols. 291–96.

——. 1913b. Über einen Versuch, die von Einstein vermutete Ablenkung des Lichtes in Gravitationsfeldern zu prüfen. *AN* 193: cols. 369–72.

——. 1913c. Zur Frage der Konstanz der Lichtgeschwindigkeit. *Phys.Z.* 14: 835–38.

——. 1913d. Anhang. *BEKSB*, Supplement, no. 15: 75–79. (Comment on Courvoisier 1913.)

——. 1914a. Über die Verschiebung der Sonnenlinien nach dem roten Ende auf

Grund der Hypothesen von Einstein und Nordström. *Phys.Z.* 15: 369–71. (See Nordström 1912.)

———. 1914b. Positionen und Eigenbewegungen von 154 Sternen. *AN* 197: cols. 297–302.

———. 1914c. Bedeckung des Sterns BD +12.2138 durch den Mond während der totalen Sonnenfinsternis am 21. August 1914. *AN* 197: cols. 335–36.

———. 1914d. Über die Verschiebung der Sonnenlinien nach dem roten Ende auf Grund der Äquivalenzhypothese von Einstein. *AN* 198: cols. 265–70.

———. 1915a. Über die Erklärung der Anomalien im Planetensystem durch die Gravitationswirkung interplanetarer Massen. *AN* 201: cols. 49–56. (Comment on von Seeliger 1906b; see also the reply in von Seeliger 1915.)

———. 1915b. Über neue Fortschritte im Dreikörperproblem. *Natw.* 3: 213–17.

———. 1915c. Die Bedeutung der Spektroskopie für die Stellarastronomie im Anschluß an W. W. Campbell: "Stellar motions." *Natw.* 3: 401–3.

———. 1915/16. Über die Gravitationsverschiebung der Spektrallinien bei Fixsternen. (**a**) Abstract in *Phys.Z.* 16: 115–17; (**b**) *AN* 202: cols. 17–24. (See also the continuation in Freundlich 1916g and 1919c; and see critique by Ludendorff 1916.)

———. 1916. *Die Grundlagen der Einsteinschen Gravitationstheorie.* (**a**) *Natw.* 4: 363–72, 386–92; (**b**) offprint (Berlin: Springer); (**c**) English trans. by Henry L. Brose (Cambridge: Cambridge University Press, 1920). (See also Cunningham 1920, Harrow 1921.)

———. 1916e. Über die Bestimmung der Solarkonstante und die dabei zutage getretenen Lichtwechsel der Sonne. *Natw.* 3: 606–9.

———. 1916f. Comments on A. S. Eddington's article on gravitation. *Obs.* 39: 70–71. (Comment on Eddington 1915.)

———. 1916g. Bemerkungen zu meinem Aufsatz in A.N. 4826. *AN* 202: cols. 147–48.

———. 1916h. Über die Dynamik der Sternhaufen. *Natw.* 4: 672.

———, ed. 1916i. *Katalog von 1886 Sternen zwischen +79° und +90° Deklination. Beobachtet von L. Courvoisier und E. Freundlich; bearbeitet von E. Freundlich.* Veröffentlichungen der Sternwarte Berlin-Babelsberg 2, no. 1.

———. 1917. Die Einsteinsche Gravitationstheorie der allgemeinen Relativitätstheorie zu den Hypothesen der klassischen Mechanik. *VAG* 52: 129–51.

———. 1918. Über die singulären Stellen der Lösungen des n-Körper-Problems, 1. Mitteilung. *SB.Berlin:* 168–88 (with announcement, p. 79). (Continuation never published.)

———. 1919a. Albert Einstein: Zum Siege seiner Relativitätstheorie. *VZ* Supplement (Nov. 30).

———. 1919b. Zur Prüfung der allgemeinen Relativitätstheorie. *Natw.* 7: 629–36; with Bemerkungen ibid.: 696.

——. 1919c. Über die Gravitationsverschiebung der Spektrallinien bei Fixsternen, II. Mitteilung. *Phys. Z.* 20: 561–70. (Continuation of Freundlich 1915/16.)

——. 1919d. Über ein Paradoxon in der Jacobischen Vorlesung über Dynamik. *AN* 208: cols. 209–12.

——. 1920. Die Entwicklung des physikalischen Weltbildes bis zur allgemeinen Relativitätstheorie. (**a**) *Die Weißen Blätter* 7: 174–91; (**b**) *Sitzungsberichte des Vereins zur Förderung des Gewerbefleißes* (Mar. 1) 99: 45–59 (with discussion; see also comment by Glaser 1920: ibid., no. 6).

——. 1920c. *Relativitätstheorie: Drei Vorträge (gehalten auf Einladung der Badischen Anilin- and Soda-Fabrik in Ludwigshafen)* (ca. 1920). (**d**) English trans. as *The theory of relativity: Three lectures for chemists*, with introduction by V. Haldane. New York: Dutton. (Inaccessible to author.)

——. 1920e. Ein Bericht der englischen Sonnenfinsternisexpedition über die Ablenkung des Lichtes im Gravitationsfelde der Sonne. *Natw.* 8: 667–73. (Review of Dyson et al. 1920.)

——. 1921. Über die Durchmesser der Fixsterne. *Natw.* 9: 191–92.

——. 1922a. Über die Träge- und Schwere-Masse der Materie und Energie. *NTNI* 82: 271–73.

——. 1922b. Über die Grundprobleme der Relativitätstheorie. *NTNI* 82: 274–75.

——. 1923a. Über die Beobachtung der Lichtablenkung während der totalen Sonnenfinsternis am 21. Sept. 1922. *Natw.* 11: 762–65.

——. 1923b. Holländisch-Deutsche Sonnenfinsternis nach Christmas Island. *AN* 218: cols. 13–16.

——. 1923c. Zur Dynamik der kugelförmigen Sternhaufen. *Phys. Z.* 24: 221–24. (See also Freundlich and Heiskanen 1923.)

——. 1923d. Über die Bedeutung der Physik des Unmeßbarkleinen für die physikalische Forschung. *Natw.* 11: 399–402.

——. 1925a. Der Michelsonsche Versuch über den Einfluß der Gravitation auf die Ausbreitung des Lichts. *Natw.* 13: 485–86.

——. 1925b. Das neue Einstein-Observatorium in Potsdam. *Sterne* 5: 33–41.

——. 1926. Thermodynamik der Gestirne. *HdP* 11: 201–37.

——. 1927a. Die Energiequelle der Sterne. *EEN* 6: 27–43. (See also Freundlich 1937a.)

——. 1927b. *Das Turmteleskop der Einstein Stiftung.* Berlin: Springer.

——. 1928a. Über die Quelle der Sonnenstrahlung. *Scientia* 44: 239–46.

——. 1928b. Die Tagung der Intern. Astron. Union in Leiden. *Natw.* 16: 730.

——. 1928c. Denkschrift über Förderung der deutschen astronomischen Forschung. *DF* 2: 152–70.

——. 1928d. Das große Turm-Teleskop in Potsdam—Ein Besuch im Einstein-Turm. *Berliner Illustrierte Zeitung* 37, no. 51: 2195–96.

——. 1929a. Über die Bedeutung der durchdringenden kosmischen Strahlung für die Astrophysik. *NTNI* 89: 21–27.

——. 1929b. Die totale Sonnenfinsternis am 9. Mai: Keine endgültigen Resultate. *Berliner Tageblatt* 58, no. 358, Aug. 1: supplement 1, p. 1.

——. 1929c. Theory supported, Einstein declares: Observers in Sumatra cable him check of star's rays near eclipse was good. *NYT*, May 11: p. 21, col. 6.

——. 1930a. Bericht über die Potsdamer Sonnenfinsternis-Expedition nach Tokengon–Nordsumatra 1929. *Natw.* 18: 313–23.

——. 1930b. Heutiger Stand des Nachweises der relativistischen Rotverschiebung der Fraunhoferschen Linien. *Natw.* 18: 513–19. (See also announcement in *Verhandlungen der Physikalischen Gesellschaft zu Berlin* 11 [1930]: 10.)

——. 1930c. The 200-inch telescope. *Natw.* 18: 707–8.

——. 1930d. Die Erforschung der Sonne. *SDM* 28: 100–104.

——. 1930e. [Short biography in *Reichshandbuch der Deutschen Gesellschaft* 1: 481.]

——. 1930f. Über die Bedeutung der totalen Sonnenfinsternisse. *DF* 12: 57–78.

——. 1931a. Review of Harlow Shapley, *Star Cluster* (Harvard Monographs, 2 [New York, 1930]). *Natw.* 19: 40.

——. 1931b. Review of Gleich 1930. *Natw.* 19: 252–53.

——. 1931c. Über die Ablenkung des Lichtes im Schwerefeld der Sonne. *FuF* 21: 292–93. (Also mentioned in *SB.Berlin*, 1931: 317.)

——. 1931d. Der innere Aufbau der Sterne nach E. A. Milne. *Natw.* 19: 685–88. (See also Freundlich 1937a.)

——. 1931/32. Die Frage nach der Endlichkeit des Weltraums als astronomisches Problem behandelt. *Erkenntnis* 2: 42–60.

——. 1932a. Über Probleme der Astrophotographie. In *Berichte über den 8. Internationalen Kongreß für wissenschaftliche und angewandte Photographie, Dresden, 1931*: 296–304. Leipzig: Barth.

——. 1932b. Recent measurements of the deflection of light by the sun's gravitational field. *Obs.* 55: 1–5.

——. 1933a. Ein neuartiger Versuch von E. A. Milne, das kosmologische Problem zu lösen und die Expansion der Spiralnebel zu deuten. *Natw.* 21: 54–59. (See Milne 1924.)

——. 1933b. Die gegenwärtige Anschauung vom Aufbau des Weltgebäudes. *Natw.* 21: 86–93.

——. 1934a. Der Aufbau des Sternsystems. In Grotrian and Kopff, eds., 1934: 168–212.

——. 1934b. Auch im Einsteinturm gilt nun der Hitler-Gruß: Weshalb Gelehrte ihre Stellung verlassen mußten. *Pariser Tageblatt*, Mar. 25. (Facsimile in Hentschel 1992c: 150.)

——. 1934/35a. Schwere des Lichtes. In *Actes du Congrès interbalkanique de mathématiciens* 1 (Athens, 1934): 143–46. (Published 1935.)

——. 1934/35b. Die Expansion der Spiralnebel und die Metrik des Weltraumes. Ibid.: 153–56.

——. 1934/35c. Über den heutigen Stand der Prüfung der Relativitätstheorie. *Sofijski Universitet Kliment Ochridski: Godisnik na Sofijkija Universitet Officialen del . . .* [Annual of the University of Sofia] 31 (math.-phys. ser.): 309–21.

——. 1935a. Das Alter der Welt und die Energiequelle der Gestirne. *Erkenntnis* 5: 323–36.

——. 1935b. Zur Theorie des "limb-Effektes" auf der Sonne. *PIUO* 1: 2–6. (**c**) In *Archives de la Société Turque des Sciences Physiques Naturelles* 1: 85–91. (**d**) English trans. (by J. Evershed) as On the theory of "limb effect" in the sun: Royal Astronomical Society, Evershed Papers, box 5. (Reply to Hunter 1934.)

——. 1935e. Entartung der Ausgleichsg. in dem Problem der Bestimmung der Lichtablenkung im Schwerefeld der Sonne. *PIUO* 1: 44–49.

——. 1936. Über Struktureigenschaften der Sternsysteme. *PIUO* 5: 85–91.

——. 1937a. O vnitrni starbe teles nebeslyeh. [On the interior constitution of the stars; in Czech.] *Rise Hvezd* 18: 25–30, 52–54, 80–81, 103–4. (See also Freundlich 1927a, 1931d.)

——. 1937b. [Textbook in astronomy; in Turkish: Istanbul. (Inaccessible to author.)]

——. 1943–59. Annual reports of the University Observatory, St Andrews (Scotland). *MNRAS* 103 (1943): 82; 105 (1945): 115; 107 (1947): 73; 108 (1948): 68–69; 109 (1949): 165–66; 110 (1950): 151–52; 111 (1951): 201; 112 (1952): 306–8; 113 (1953): 332–33; 114 (1954): 324–25; 115 (1955): 157–58; 116 (1956): 188–89; 117 (1957): 285–86; 118 (1958): 337–38; 119 (1959): 380–81.

——. 1945a. The structure of the globular star clusters. *MNRAS* 105: 237–43.

——. 1945b. *Air navigation.* Edinburgh: Oliver and Boyd.

——. 1947. On the theory of globular star clusters. *MNRAS* 107: 268–73.

——. 1950a. On the measurement of the relativistic light deflection. *Astronomical Journal* 55: 245–47. (Reply to van Biesbroek 1950; see also his response: ibid. 247.)

——. 1950b. Schmidt-Cassegrain telescope at Dundee. *Nature* 165: 703–4.

——. 1950c. The development of the astronomical telescope. *Journal of Scientific Instruments* 27: 233–37.

——. 1951a. *Cosmology.* Chicago: University of Chicago Press. (*International Encyclopedia of Unified Science,* vol. 1, no. 8.)

——. 1951b. Albert von Brunn †. *AN* 280: col. 142.

——. 1952a. Über den gegenwärtigen Stand der Prüfung der allgemeinen Relativitätstheorie. *Sterne* 28: 220–22. (Summary by E. Ahnert and R. Ahnert.)

——. 1952b. Der gegenwärtige Stand der Prüfungen der allgemeinen Relativitätstheorie. *Wiss.Ann.* 1: 35–42.

——. 1953a. Über die Rotverschiebung der Spektrallinien. *Nachrichten der Akademie der Wissenschaften zu Göttingen*, Math.-phys. Klasse, ser. 2a, no. 7: 96–102. (**b**) In Contributions of the Observatory of St Andrews, no. 4. (See also reply in Born 1954.)

——. 1953c. E. von der Pahlen †. *Natw.* 40: 589.

——. 1953d. The educational value of the study of astronomy. *Alumnus Chronicle* (University of St Andrews) 40: 2–14.

——. 1953e. Die Entwicklung des astronomischen Fernrohrs. *Wiss.Ann.* 2: 657–68. (Reprinted in Contributions of the Observatory of St Andrews, no. 13.)

——. 1954a. Red shifts in the spectra of celestial bodies. *Phil.Mag.*, 7th ser., 45: 303–19. (**b**) Abstract in *PPSL* A67: 192–93. (**c**) Short review in *Phys.Rev.*, 2d ser., 95: 654. (**d**) In Contributions of the Observatory of St Andrews, no. 5. (See also replies in Born 1954, Burbidge 1954, ter Haar 1954, and McCrea 1954.)

——. 1954e. The general red shift of spectral lines in the spectra of celestial bodies. *Lunds Universitets Årsskrift*, n.s., 50: 106.

——. 1954f. Über Rotverschiebungen der Spektrallinien kosmischer Lichtquellen. *FuF* 28: 353–57.

——. 1954g. Zum Gedächtnis. Prof. Dr. phil. Walter Grotrian, ordentl. Mitglied der Deutschen Akademie der Wissenschaften zu Berlin. *Wiss.Ann.* 3: 439–41.

——. 1955a. On the empirical foundation of the general theory of relativity. *ViA* 1: 239–46.

——. 1955b. Wird Einsteins Voraussage in Ceylon bestätigt werden? *Neue Zeitung*, Jan. 9, no. 7: p. 16, cols. 1–5.

——. 1955c. Probleme und gegenwärtiger Stand der empirischen Begründung der allgemeinen Relativitätstheorie. Transcript of interview broadcast, Rundfunk im Amerikanischen Sektor von Berlin Funk-Univ., Apr. 19, 1955.

——. 1957. Du déplacement général vers le rouge des raies du spectre solaire. *Annales de Physique*, 13th ser., 2: 765–77.

——. 1958. *Celestial mechanics*. London: Pergamon Press.

——. 1959. The empirical foundation of the general theory of relativity. *Scientia* 94 (annus 53): 181–87.

——. 1960a. Der Nachweis der Schwere des Lichtes. *Natw.* 47: 123–27.

——. 1960b. *Pourquoi n'a-t-on pas encore prouvé l'effet relativiste de déplacement vers le rouge*. Séminaire JANET (Mécanique Analytique et Mécanique Céleste), année 4 (1960/61), no. 2 (Oct. 16, 1960).

——. 1961. Über eine allgemeine Rotverschiebung der Fraunhoferschen Linien. *Natw.* 48: 709–11.

——. 1964. Betrachtungen zu dem Problem der Schwere des Lichtes. *ZAP* 58: 283–87.

——. 1969. Wie es dazu kam, daß ich den Einsteinturm errichtete. *PB* 25: 538–41.

Freundlich, E. F., and A. von Brunn. 1933. Über die Theorie des Versuchs der Bestimmung der Lichtablenkung im Schwerefeld der Sonne. *ZAP* 6: 218–35.

Freundlich, E. F., et al. 1930. [E. F. Freundlich, A. von Brunn, and H. Brück.] Über den Verlauf der Wellenlängen der Fraunhoferschen Linien längs der Sonnenoberfläche. *ZAP* 1: 43–57. (See also von Brunn 1930.)

——. 1927. [E.F. Freundlich, E. Hopf, and U. Wegener.] On the integral equation for radiative equilibrium. *MNRAS* 88: 139–42. (Reply to Parchomenko 1926.)

——. 1931a. [E. F. Freundlich, H. von Klüber, and A. von Brunn.] Ergebnisse der Potsdamer Expedition zur Beobachtung der Sonnenfinsternis von 1929, Mai 9, in Takengon (Nordsumatra), 5. Mitteilung: Über die Ablenkung des Lichtes im Schwerefeld der Sonne. *ZAP* 3: 171–98. (See comment in Trümpler 1932b.)

——. 1931b. Die Ablenkung des Lichtes im Schwerefeld der Sonne. *Abh.Berlin*, no. 1: 1–61, with plates. (See comment by Jackson 1931.)

——. 1931c. [E. F. Freundlich, C. E. St. John, and A. Einstein.] New proofs found of an Einstein idea: He is told findings show rays of star light bend in passing near sun. *NYT*, Feb. 21: p. 3, col. 3.

——. 1932a. [E. F. Freundlich, H. von Klüber, and A. von Brunn.] Bemerkung zu Herrn Trümplers Kritik. *ZAP* 4: 221–23. (Reply to Trümpler 1932b.)

——. 1932b. Bemerkung zu Herrn Ludendorffs Ausführungen in Nr. 5848 dieser Zeitschrift. *AN* 244: col. 415. (Comment on Ludendorff 1932a.)

——. 1933. Weitere Untersuchungen über die Bestimmung der Lichtablenkung im Schwerefeld der Sonne. *Annalen van de Bosscha-Sterrenwacht Lembang* 5: B1–B64.

Freundlich, E. F., and E. G. Forbes. 1956. On the red shift of solar lines. Parts 1 and 2. *AdA* 19: 183–98, 215–22.

——. 1959. On the red shift of solar lines. Part 3. *AdA* 22: 727–32.

Freundlich, E. F., and W. Gleissberg. 1935. Zur Frage der Entartung der Ausgleichung in dem Problem der Bestimmung der Lichtablenkung im Schwerefeld der Sonne. *PIUO*, n.s., 1, no. 2: 44–49.

Freundlich, E. F., and W. Heiskanen. 1923. Über die Verteilung der Sterne verschiedener Masse in den kugelförmigen Sternhaufen. *Z.Phys.* 14: 226–39.

Freundlich, E. F., and W. Herrmann. 1931. Beschreibung eines großen Plattenmeßapparates. *ZfI* 51: 582–86.

Freundlich, E. F., and E. Hochheim. 1924. Über den Ursprung der sog. Cyanbande bei 3883 Å. *Z.Phys.* 26: 102–5.

Freundlich, E. F., and K. Kühne. 1928. Über die Messung von Sterndurchmessern nach Pokrowski. *AN* 233: cols. 271–74.

Freundlich, E. F., and R. Kurth. 1955. Über die Grundlagen, Ziele und Mög-
lichkeiten einer mechanischen Theorie der Sternsysteme. *Natw.* 42: 167–69.

Freundlich, E. F., and W. Ledermann. 1944. The problem of the accurate determination of the relativistic light deflection. *MNRAS* 104: 40–47.

Freundlich, E. F., and E. von der Pahlen. 1923. Untersuchung des K-Effektes auf
Grund des Katalogs von Radialgeschwindigkeiten von J. Voute. *AN* 218:
cols. 369–400.

Fritze, Klaus. 1976. Zur Geschichte der Sternwarte Babelsberg. *Wissenschaft und
Fortschritt* (Berlin) 26: 299–302.

Fujita, Yoshio. 1934. On the prism spectrograph of the tower telescope in the
Tokyo Astronomical Observatory. *Proceedings of the Physical-Mathematical Society of Japan*, 3d ser., 16: 327–31.

Fürst, Artur. 1926. Der Sternenstrahl im Keller: Ein Besuch im Einstein-Turm
zu Potsdam. *Koralle* 2, no. 11: 28–37.

Fürth, R., and E. F. Freundlich. 1955. On a possible improvement of Michelson's
method for the determination of stellar diameters in poor visibility. *ViA* 1:
395–400.

Galison, Peter. 1993. The cultural meaning of *Aufbau*. In F. Stadler, ed., *Scientific
philosophy: Origins and developments*: 75–93. Dordrecht: Kluwer.

Gay, Peter. 1968. *Weimar culture: The outsider as insider*. New York: Harper and
Row.

Geertz, Clifford. 1973. *The interpretation of cultures: Selected essays*. New York:
Basic Books.

Giedion, Siegfried. 1941. *Space, time and architecture: The growth of an idea*. Cambridge, Mass.: Harvard University Press. (**a**) 1st ed. (1941); (**b**) 5th ed. (1967;
see esp. p. 436 on Einstein).

Gingerich, Owen. 1984. *Astrophysics and twentieth-century astronomy to 1950*. Cambridge: Cambridge University Press.

Glaser, Ludwig C. 1920. Bemerkungen zu dem im Verein zur Beförderung des
Gewerbefleißes am 1. März 1920 von Herrn Dr. Freundlich gehaltenen
Vortrage "Die Entwicklung des physikalischen Weltbildes bis zur allge-
meinen Relativitätstheorie." *Sitzungsberichte des Vereins zur Beförderung des
Gewerbefleißes* 99, no. 6: 105–6. (Comment on Freundlich 1920b.)

von Gleich, Gerold. 1930. *Einsteins Relativitätstheorie und physikalische Wirklich-
keit*. Leipzig: Barth. (See also review in Freundlich 1931b.)

Gleissberg, Wolfgang. 1967. Astronomie in der Türkei. *Sterne und Weltraum* 6:
275–79.

Glick, Thomas F., ed. 1987. *The comparative reception of relativity*. Dordrecht:
Reidel.

Gollnow, H. 1949. Turmteleskope. *Natw.* 36: 175–82, 213–17.

Gondolatsch, F. 1961. August Kopff. *MAG*: 5–16.

Gössel, Peter, and Gabriele Leuthäuser. 1990. *Architektur des 20. Jahrhunderts*.
Cologne: Taschen. (See esp. pp. 118–35 on expressionist architecture.)

Grebe, Leonhard C. F., and Albert Bachem. 1919. Über den Einsteineffekt im Gravitationsfeld der Sonne. *Verhandlungen der Deutschen Physikalischen Gesellschaft* 21: 454–64.

Grotrian, Walter. 1931. Ergebnisse der Potsdamer Expedition zur Beobachtung der Sonnenfinsternis von 1929, Mai 9, in Takengon (Nordsumatra). (a) 1. Mitteilung: Spektroskopische Untersuchungen an Korona und Protuberanzen. *ZAP* 2: 106–32. (b) 6. Mitteilung: Über die Intensitätsverteilung des kontinuierlichen Spektrums der inneren Korona. *ZAP* 3: 199–226.

———. 1933. Die Beobachtung der Sonnenkorona außerhalb totaler Sonnenfinsternisse. *Sterne* 13: 73–80.

———. 1934. Über die physikalische Natur der Sonnenkorona. *Sterne* 14: 145–57.

———. 1939a. Zur Frage der Deutung der Linien im Spektrum der Sonnenkorona. *Natw.* 27: 214.

———. 1939b. Sonne und Ionosphäre. *Natw.* 27: 555–63, 569–77.

———. 1952. 30 Jahre Forschungsarbeit im Einsteinturm in Potsdam. *Wiss.Ann.* 1: 79–93.

Grotrian, W., and A. Kopff, eds. 1934. *Zur Erforschung des Weltalls: Acht Vorträge über Probleme der Astronomie und Astrophysik*. Berlin: Springer.

Gußmann, E. A. 1975. 100 Jahre Astrophysikalisches Observatorium Potsdam— Arbeiten zur Theorie der Sternspektren. *Sterne* 51: 219–27. (See also Scholz 1975, Wempe 1975.)

Guthnick, Paul. 1941. Hans Ludendorff zum Gedächtnis. *FuF* 17: 350–51.

———. 1942. Hans Ludendorff. *VAG* 77: 2–15.

ter Haar, Dirk. 1954. On some more remarks on Freundlich's red-shift. *Phil.Mag.*, 7th ser., 45: 1023–24. (Comment on Freundlich 1954a and McCrea 1954; see also Burbidge 1954, Kirschstein 1954.)

Hajos, E. M. 1929. Berliner Architektur und Architekten von heute. *Kunstwanderer*: 493–97.

Hale, George Ellery. 1905. The solar observatory of the Carnegie Institution of Washington. *ApJ.* 21: 151–72.

———. 1908. The tower telescope of the Mount Wilson Solar Observatory. *ApJ.* 27: 204–12. (Contributions from the Mt. Wilson Solar Observatory, no. 23.)

———. 1915. *Ten years' work of a mountain observatory. A brief account of the Mount Wilson Solar Observatory of the Carnegie Institution of Washington*. Publications of the Carnegie Institution of Washington, vol. 235. Washington, D.C.

Hale, George Ellery, and Seth B. Nicholson. 1938. *Magnetic observations of sunspots, 1917–1924*. Publications of the Carnegie Institution of Washington, vol. 498. Washington, D.C.

Harrington, Anne. 1991. Interwar "German" psychobiology: Between nationalism and the irrational. *Science in Context* 4: 429–41.

Harrow, Benjamin. 1921. Reviews of Eddington 1920, Freundlich 1916c, and Einstein 1920. *NYT*, Apr. 17: sec. 3, p. 3, cols. 1–5; and p. 30.

Hassenstein, Walter. 1941. Das Astrophysikalische Observatorium Potsdam in den Jahren 1875–1939. *Mitteilungen des Astrophysikalischen Observatoriums Potsdam* 1: 1–56.

Hearnshaw, J. B. 1986. *The analysis of starlight: One hundred and fifty years of astronomical spectroscopy*. Cambridge: Cambridge University Press.

——. 1992. Doppler and Vogel—Two notable anniversaries in stellar astronomy. *ViA* 35: 157–77.

Heilbron, John. 1986. *The dilemmas of an upright man: Max Planck as spokesman for German science*. Berkeley: University of California Press.

Heisenberg, Werner. 1971. Das Kaiser-Wilhelm-Institut für Physik: Geschichte eines Instituts. *Jahrbuch der Max-Planck-Gesellschaft*: 46–89.

Hellwag, Fritz. 1925. Einstein-Turm. *Dekorative Kunst* 29 (*Kunst: Monatshefte für freie und angewandte Kunst* 54): 156–60.

Henderson, Linda Dalrymple. 1983. *The fourth dimension and non-Euclidean geometry in modern art*. Princeton: Princeton University Press.

Hentschel, Klaus. 1985. Review of Pyenson 1985. *AIHS* 35: 496–98.

——. 1990a. *Interpretationen und Fehlinterpretationen der speziellen und allgemeinen Relativitätstheorie durch Zeitgenossen Albert Einsteins*. Basel: Birkhäuser. (Science Networks, vol. 6.)

——. 1990b. *Die Korrespondenz Petzoldt-Reichenbach: Zur Entwicklung der "wissenschaftlichen Philosophie" in Berlin*. Berlin: Sigma.

——. 1991. *Julius und die anomale Dispersion—Facetten der Geschichte eines gescheiterten Forschungsprogrammes*. Studien aus dem Philosophischen Seminar (Hamburg), ser. 3, no. 6.

——. 1992a. Grebe/Bachems photometrische Analyse der Linienprofile und die Gravitations-Rotverschiebung, 1919–1922. *Annals of Science* 49: 21–46.

——. 1992b. Einstein's attitude towards experiments: Testing relativity theory, 1907–1927. *Studies in History and Philosophy of Science* 23: 593–624.

——. 1992c. *Der Einstein-Turm: Erwin F. Freundlich und die Relativitätstheorie—Ansätze zu einer 'dichten Beschreibung' von institutionellen, biographischen und theoriengeschichtlichen Aspekten*. Berlin: Spektrum.

——. 1993a. The conversion of St. John: A case study on the interplay of theory and experiment. *Science in Context* 6, no. 1: 137–94.

——. 1993b. The discovery of the redshift of solar spectral lines by Rowland and Jewell in Baltimore around 1890. *HSPS* part 23, 2: 219–78.

——. 1994. Erwin Finlay Freundlich and testing Einstein's theory of relativity. *Archive for the History of the Exact Sciences* 47, no. 2: 143–201.

——. 1995a. Einstein—Freundlich—Mendelsohn. *Sterne* 71: 151–67.

——. 1995b. Zum Zusammenspiel von Instrument, Experiment und Theorie am Beispiel der Rotverschiebung im Sonnenspektrum und verwandter spektraler Verschiebungseffekte von ca. 1880 bis etwa 1960. Habilitation thesis, Hamburg University, submitted May 8, 1995.

——, ed. 1996. *Physics and national socialism: An anthology of primary sources*. Basel: Birkhäuser.

Herf, Jeffrey. 1984. *Reactionary modernism: Technology, culture and politics in Weimar and the Third Reich*. Cambridge: Cambridge University Press.

Hermann, Armin, ed. 1968. *Albert Einstein / Arnold Sommerfeld Briefwechsel*. Basel: Schwabe.

——. 1973. Laue, Max von. *DSB* 8: 50–53.

——. 1989. *Nur der Name war geblieben: Die abenteuerliche Geschichte der Firma Carl Zeiss*. Stuttgart: DVA.

Herneck, Friedrich. 1979. *Max von Laue*. Leipzig: Teubner.

Herrmann, Dieter B. 1973. Zur Vorgeschichte der Astrophysik in Deutschland und in den USA: Ein quantitativer Vergleich. *NTM* 10: 38–44.

——. 1974. Müller, Gustav. *DSB* 9: 563–64.

——. 1975. Zur Vorgeschichte des Astrophysikalischen Observatoriums, Potsdam (1865–1874). *AN* 296: 245–59.

——. 1976. Vogel, Hermann Carl. *DSB* 14: 54–57.

——. 1981a. Karl Friedrich Zöllner und sein Einfluß auf das Forschungsprogramm des Astrophysikalischen Observatoriums Potsdam. *Berliner Wissenschaftshistorische Kolloquien* 22: 49–54.

——. 1981b. Das Astrophysikalische Observatorium Potsdam, oder: Vom Zeitfaktor in der Wissenschaftsgeschichte. Ibid. 24: 196–204.

——. 1984. *Geschichte der modernen Astronomie*. Cologne: Aulis.

Herrmann, D. B., and Jürgen Hamel. 1975. Zur Frühentwicklung der Astrophysik: Das internationale Forscherkollektiv, 1865–1899. *NTM* 12: 25–30.

Herrmann, W., and E. F. Freundlich. 1931. Beschreibung eines großen Plattenmeßapparates. *ZfI* 51: 582–86.

Hertzsprung, Ejnar. 1917. Karl Schwarzschild. *APJ*. 45: 285–92.

Hetherington, Norris S. 1982. Philosophical values and observations in Edwin Hubble's choice of a model of the universe. *HSPS* 13: 41–67.

Hoch, Paul K. 1983. The reception of Central European refugee physicists of the 1930's: U.S.S.R., U.K., U.S.A. *Annals of Science* 40: 217–46.

Hoffmann, Dieter. 1991. Erwin Finlay-Freundlich (1885–1964) und die Profilierung der astrophysikalisch-astronomischen Forschung in der Tschechoslowakei. *Zbornik Cesdef '91*: 137–51.

Hopmann, Josef. 1922. Die Sonnenfinsternis-Expedition: Zur Prüfung der Relativitätstheorie. *VZ* no. 269, June 9, morning ed., suppl. 1.

——. 1923. Die Deutung der Ergebnisse der amerikanischen Einstein-Expedition. *Phys. Z.* 24: 476–85.

——. 1954. A possible test for Freundlich's redshift. *Obs.* 74: 169.

Howard, Don, and John Stachel, eds. 1989. *Einstein and the history of general relativity*. Basel: Birkhäuser. (Einstein Studies, 1.)

Hüter, Karl-Heinz. 1987. *Architektur in Berlin 1900–1933*. Dresden: VEB Verlag der Kunst. (See esp. pp. 94–106, 321–55.)

Hufbauer, Karl. 1991. *Exploring the sun: Solar science since Galileo*. Baltimore: Johns Hopkins University Press.

———. 1993. Breakthrough on the periphery: Bengt Edlén and the identification of the coronal lines, 1939–45. In Svante Lindqvist, ed., *Center on the periphery: Historical aspects of twentieth-century Swedish physics*: 199–237. Canton, Mass.: Watson Publishing International.

Hughes, Thomas. 1986. The seamless web: Technology, science, etcetera, etcetera. *Social Studies of Science* 16: 281–92.

Humphreys, William Jackson. 1908. Bericht über die Verschiebung von Spektrallinien durch Druck. *JRE* 5: 324–74.

Hunter, Alan. 1934. The solar limb effect. *MNRAS* 94: 594–603. (See also reply in Freundlich 1935d.)

Isaksson, Eva. 1985. Der finnische Physiker Gunnar Nordström und sein Beitrag zur Entstehung der allgemeinen Relativitätstheorie A. Einsteins. *NTM* 22: 29–52.

Jack, D. 1951. The Schmidt-Cassegrain telescope. *Alumnus Chronicle* (University of St Andrews) 36: 23–28. (This article is often wrongly attributed to Freundlich.)

Jackson, J. 1931. The deflection of light in the sun's gravitational field. *Obs.* 54: 292–96. (Critique of Freundlich et al. 1931b.)

Jäger, Friedrich Wilhelm. 1961. Paul ten Bruggencate. *MAG* 15: 21–28.

———. 1986. Der Einstein-Turm in Potsdam und die Relativitätstheorie. *Sterne* 62: 14–23.

James, Kathleen. 1994. Expressionism, relativity, and the Einstein Tower. *Journal of the Society of Architectural Historians* 53, no. 4: 392–413.

Joedicke, Jürgen. 1966. *A history of modern architecture*. New York: Praeger. (See esp. p. 65 on the Einstein Tower.)

Joël, Kurt. 1923. Deutsche Forscher nach Mexiko: Zur Sonnenfinsternis am 10. Sept. *VZ*, July 6.

Julius, Willem Henri. 1910. Les raies de Fraunhofer et la dispersion anomale de la lumière. *Radium* 7: 281–84.

Kienle, Hans. 1924. Kosmische Refraktion. *Phys.Z.* 25: 1–6. (See also Kopff 1924.)

———. 1925. Hugo von Seeliger. *Natw.* 13: 613–19.

———. 1955. Walter Grotrian. *MAG*: 5–9.

———. 1962. Paul ten Bruggencate. *Natw.* 49: 73–74.

Kiepenheuer, K. O. 1946/48. Solar-terrestrische Erscheinungen. In *Fiat— Review of German sciences: Astronomy, astrophysics and cosmogony*: 230–84. Wiesbaden: Klemm.

Kirchberger, Paul. 1925. Der Einstein-Turm. *Frankfurter Zeitung* 70, no. 904, Dec. 4, evening edition.

Kirschstein, G. 1954. Zur Rotverschiebung der Sternspektren. *PB* 10: 376. (Comment on Freundlich 1954 and ter Haar 1954.)

Kirsten, Christa, and Hans-Günther Körber, eds. 1975. *Physiker über Physiker:*

Wahlvorschläge zur Aufnahme von Physikern in die Berliner Akademie, 1870–1929.
Berlin: Akademie-Verlag.

Kirsten, Christa, and Hans-Jürgen Treder, eds. 1979. *Albert Einstein in Berlin, 1913–1933.* 2 vols. Berlin: Akademie-Verlag.

Klein, M. J., et al., eds. 1993a. [M. J. Klein, A. J. Kox, J. Renn, and R. Schulmann.] *The collected papers of Albert Einstein*, vol. 3: *The Swiss years: Writings, 1909–1911.* Princeton: Princeton University Press. (**b**) English trans. by A. Beck, ibid.

——. 1993c. [M. J. Klein, A. J. Kox, and R. Schulmann.] *The collected papers of Albert Einstein*, vol. 5: *The Swiss years: Correspondence, 1902–1914.* Princeton: Princeton University Press. (**d**) English trans. by A. Beck, ibid. (1995).

von Klüber, Harald. 1926. Die Deutsch-Niederländisch-Indische Sonnenfinsternis-Expedition nach Sumatra 1925/26. *Sterne* 6: 157–64.

——. 1927a. Quantitative Untersuchungen an Absorptionslinien im Sonnenspektrum. *Z.Phys.* 44: 481–516.

——. 1927b. Photometrie von Absorptionslinien im Sonnenspektrum. *AN* 231: cols. 417–24.

——. 1929. Ausfahrt zur Sonnenfinsternis. *Sterne* 9: 85–89.

——. 1931. Ergebnisse der Potsdamer Sonnenfinsternisexpedition 1929, Mai 9.: Photographische Photometrie der Sonnenkorona. (**a**) 2. Mitteilung. *ZAP* 2: 289–328. (**b**) 4. Mitteilung. Ibid. 3: 142–62.

——. 1931c. Die astronomische Nachprüfung der Lichtablenkung im Schwerefeld der Sonne. *Sterne* 11: 141–55.

——. 1932. Über eine Horizontalkamera mit Präzisions-Coelostaten für astronomische Beobachtungen. *ZfI* 52: 381–93.

——. 1936. Kalorimetrie. *HdA* 7: 71–83.

——. 1944/48a. Zum Nachweis schwacher Magnetfelder auf der Sonnenoberfläche. *ZAP* 24: 1–21.

——. 1944/48b. Über den Nachweis und die Messung lokaler Magnetfelder auf der Sonnenoberfläche. *ZAP* 24: 150–80.

——. 1955. Die Sonnenfinsternis vom 30. Juni. *Natw.* 42: 3–8.

——. 1960. The determination of Einstein's light-deflection in the gravitational field of the sun. *ViA* 3: 47–77.

——. 1965. Erwin Finlay-Freundlich †. (**a**) *AN* 288: 281–86; (**b**) abstract in *QJRAS* 6: 82–84.

Knobloch, Eberhard, and Burghard Weiss. 1987. Astronomen und Astrophysiker in Berlin. In W. Treue and G. Hildebrandt, eds., *Berlinische Lebensbilder*, vol. 1: *Naturwissenschaftler:* 79–90. Berlin: Colloquium.

König, G. K. 1966. Mendelsohn e l'Einsteinturm. (**a**) *Casabella* 303: 40–45. (**b**) English trans. (by L. Mendelsohn) as Mendelsohn and the Einsteinturm. At KB (Berlin), Mendelsohn Papers, call no. I/f/6.

Kopal, Zdeněk. 1964. Prof. E. F. Freundlich. *Nature* 204: 727–28.

Kopff, August. 1924. Courvoisier-Effekt und Einstein-Effekt. *Phys.Z.* 25: 95–96. (Comment on Kienle 1924.)

———. 1941. Gedächtnisrede auf Hans Ludendorff. *Jahrbuch der Preußischen Akademie der Wissenschaften*: 225–33.

Köster, Hein. 1974. Der Einstein-Turm. *Form und Zweck* 6, no. 11: 44–46.

Krafft, Fritz. 1981. Astrophysik contra Astronomie: Das Zurückdrängen einer alten Disziplin durch die Begründung einer neuen. *Berichte zur Wissenschaftsgeschichte* 4: 89–110.

Kratzer, A. 1921. Die Termdarstellung der Cyanbanden. *Phys.Z.* 22: 552–55.

Krausse, Joachim. 1989. Vom Einsteinturm zum Wunder von Jena: Einsteins Weltbild und die Architektur. In Thomas Neumann, ed., *Albert Einstein*: 58–67. Berlin: Elephanten.

Kruse, W. 1925. Der Einstein-Turm. *Illustrierte Technik für Jedermann* 3, no. 7: 100–102.

Kühne, Günther. 1994. Erich (Eric) Mendelsohn. *Neue Deutsche Biographie* 17: 42–43.

Kuiper, Gerard P. 1946. German astronomy during the war. *Pop.Astr.* 54: 263–87.

Künzel, H. 1986. Beobachtung solarer Magnetfelder am Sonnenobservatorium Einsteinturm. *Sterne* 62: 208–17.

Lane, Barbara Miller. 1968. *Architecture and politics in Germany, 1918–1945*. Cambridge, Mass.: Harvard University Press (**a**) 1st ed. (1968); (**b**) 2nd ed. (1985; see esp. pp. 53–55 on the Einstein Tower).

Lankford, John. 1981. Amateurs and astrophysics. *Social Stds. of Science* 11: 275–303.

von Laue, Max. 1917. Nordströmsche Gravitationstheorie. *JRE* 14: 263–313.

Limberg, Jörg, and Jürgen Staude. 1994. *Erich Mendelsohns Einsteinturm in Potsdam*. Potsdam: Potsdamer Verlagsbuchhandlung. (Arbeitshefte des Brandenburgischen Landesamtes für Denkmalpflege, no. 5.)

Lippmann, G. 1895. Sur un coelostat, ou appareil à miroir, donnant une image du ciel immobile par rapport à la terre. *CRAS* 120: 1015–19.

Litten, Freddy. 1992. *Astronomie in Bayern, 1914–1945*. Stuttgart: Steiner. (Boethius, 30.)

Littmann, Mark, and Ken Willcox. 1991. *Totality: Eclipses of the sun*. Honolulu: University of Hawaii Press.

Ludendorff, Hans. 1897. *Die Jupiter-Störungen der kleinen Planeten vom Hecuba-Typus*. Berlin: Mayer and Müller. (Ph.D. thesis, University of Berlin.)

———. 1916. Bemerkungen über die Radialgeschwindigkeiten der Helium-Sterne. *AN* 202: cols. 75–84. (Reply to Freundlich 1915/16a; see also von Seeliger 1916.)

———. 1920a. Weitere Untersuchungen über die Massen der spektroskopischen Doppelsterne. *AN* 211: cols. 105–20.

———. 1920b. Über die Verschiebung der Emissionslinien bei den Mira-Sternen. *AN* 212: cols. 483–84.

——, ed. 1921. *Newcomb-Engelmanns populäre Astronomie*. 6th ed. Leipzig: Engelmann. (Ed. Ludendorff "together with Prof. Eberhard, Dr. Freundlich and Dr. Kohlschütter"; see esp. Freundlich's contributions in the sections: Die Grundgesetze der Mechanik und ihre Entwicklung seit Newton, pp. 60–76; Das Dreikörperproblem und die Bewegung der Planeten, pp. 76–83; Die Sonne, pp. 271–338; Neue Sterne, pp. 662–78.)

——. 1923a. Nikolaus Coppernicus. *Sterne* 3: 46–49.

——. 1923b. Antrittsrede. *SB.Berlin*: lxxxiv.

——. 1923/24. Vorläufiger Bericht über die deutsche Sonnenfinsternis-Expedition nach Mexiko. *AN* 220: cols. 77–78.

——. 1925a. Gustav Müller. *VAG* 60: 124–74. (b) Abstract in *AN* 225: cols. 199–200. (See also Müller 1925.)

——. 1925c. Spektralphotometrische Untersuchungen über die Sonnenkorona. *SB.Berlin*: 83–113.

——. 1928. Über die Abhängigkeit der Form der Sonnenkorona von der Sonnenfleckenhäufigkeit. *SB.Berlin*: 185–214.

——. 1930. Die astrophysikalische Expedition nach Bolivia. *DF* 12: 16–32.

——. 1931. Ludendorff, Hans. *Reichshandbuch der Deutschen Gesellschaft* 2: 1161–62.

——. 1932a. Über die Ablenkung des Lichtes im Schwerefeld der Sonne. *AN* 244: cols. 321–30. (See also reply in Freundlich et al. 1932b, and response in Ludendorff 1932b.)

——. 1932b. Erwiderung auf die vorstehende Bemerkung. *AN* 244: col. 415.

——. 1936. Die veränderlichen Sterne. *HdA* 7: 614–70.

——. 1942. *Zur Frühgeschichte der Astronomie in Berlin*. Berlin: de Gruyter. (Vorträge und Schriften der Preußischen Akademie der Wissenschaften, 9.)

Macrakis, Kristie. 1986. Wissenschaftsförderung durch die Rockefeller-Stiftung im "Dritten Reich." *Geschichte und Gesellschaft* 12: 348–79.

Mattig, Wolfgang. 1956. Die Lichtablenkung im Schwerefeld der Sonne: Bericht über die Sonnenfinsternisexpeditionen 1954 und 1955. *Sterne* 32: 177–88.

——. 1995. Von der Rotverschiebung zum solaren Magnetfeld. *Sterne* 71: 127–41.

Matukuma, T. 1940. Derivation of Einstein effect from eclipse observations of 1936. *Nature* 265: 264–65.

McCrea, William H. 1951. Edward Arthur Milne. *ONFRS* 7: 421–43.

——. 1954. Astrophysical considerations regarding Freundlich's red-shift. *Phil.Mag.*, 7th ser., 45: 1010–18. (Comment on Freundlich 1954; see also ter Haar 1954.)

Mehrtens, Herbert. 1990. *Moderne Sprache Mathematik*. Frankfurt: Suhrkamp.

Melvin, M. A. 1955. Freundlich's red-shift formula. *Phys.Rev.*, 2d ser., 98: 884–87.

Mendelsohn, Erich. 1924. Bauten und Skizzen. *Wasmuths Monatshefte für Baukunst* 8: 3–66.

———. 1930. *Erich Mendelsohn: Das Gesamtschaffen des Architekten—Skizzen, Entwürfe, Bauten.* Berlin: Mosse. (Reprinted Braunschweig: Vieweg, 1989.)

Meyer, F. 1930. Über die Entwicklung der astronomischen Instrumente im Zeisswerke Jena. *ZfI* 50: 58–99.

Meyer, Peter. 1928. *Moderne Architektur und Tradition.* Zurich: Girsberger.

Mie, Gustav. 1912/13. Grundlagen einer Theorie der Materie. *Ann.Phys.*, 4th ser., 37: 511–34; 39: 1–40; 40: 1–60.

———. 1914. Bemerkungen zur Einsteinschen Gravitationstheorie. *Phys.Z.* 15: 105–22, 169–96.

Mikhailov, A. A. 1940. Measurement of the deflection of light by the sun's gravitational field during the eclipse of June 19, 1936. *Doklady Akademii Nauk, USSR* 29: 189–90.

———. 1959. The deflection of light by the gravitational field of the sun. *MNRAS* 119: 593–608.

Milne, Edward Arthur. 1924. Recent work in stellar physics. *PPSL* 36: 94–113. (See also Freundlich 1933a.)

Moyer, Donald. 1979. Revolutions in science: The 1919 eclipse test of general relativity. In Arnold Perlmutter and L. F. Scott, eds., *On the Path of Albert Einstein*: 55–101. New York: Plenum.

Müller, Peter. 1991. Die beiden Observatorien in Potsdam und Babelsberg. *Sterne und Weltraum* 67: 550–51.

Müller, Rolf. 1925. [Bibliography of Gustav Müller.] *VAG* 60: 174–77. (See also Ludendorff 1925a.)

Münch, W. 1941. Hans Ludendorff †. *AN* 271: 294–96.

Neumark, Fritz. 1980. *Zuflucht am Bosporus: Deutsche Gelehrte, Politiker und Künstler in der Emigration 1933–1953.* Frankfurt: Knecht. (See esp. p. 100 on Freundlich.)

Neutra, Richard, and Dione Neutra. 1986. *Richard Neutra: Promise and fulfillment, 1919–1932—Selections from the letters and diaries of Richard and Dione Neutra.* Comp. and trans. Dione Neutra. Carbondale: Southern Illinois University Press.

N.N. 1921a. Der Potsdamer Astronomenkongreß. *Berliner Tageblatt* 50, no. 397 (B197), Aug. 24: p. 3.

———. 1921b. Der Potsdamer Astronomentag. Ibid., Aug. 26.

———. 1921c. Der neue Einstein-Turm auf dem Telegraphenberg bei Potsdam. *Berliner Illustrierte Zeitung*, Sept. 4 (cover).

———. 1922. Die Einstein-Expedition nach der Weihnachtsinsel: Nachprüfung der Relativitätstheorie. *Berliner Tageblatt*, no. 313, July 18: supplement 1.

———. 1926. Der Einstein-Turm. *Germania*, ser. 56, no. 464 (B264), Oct. 6, morning ed.: p. 5.

———. 1928. A new 18-inch coelostat. *Nature* 121: 247–49.

———. 1952. Nachruf (auf Dr. phil. Emmanuel von der Pahlen). *ZAP* 31: 97–98.

———. 1959. Astronomy at St Andrews: Prof. E. F. Freundlich. *Nature* 184: 768.

———. 1963. Technische Formen altern schnell—Vor zehn Jahren starb Erich Mendelsohn. *Welt*, Sept. 14.

———. 1989. Wiesbadener war ein Weggefährte Einsteins. *Allgemeine Zeitung Mainz*, July 22: p. 37.

Nordström, Gunnar. 1912. Relativitätsprinzip und Gravitation. *Phys.Z.* 13: 1126–29. (See also Freundlich 1914a.)

North, John D. 1965. *The measure of the universe: A history of modern cosmology.* Oxford: Oxford University Press. (Reprinted New York: Dover, 1990.)

———. 1992. Einstein, Nordström and the early demise of scalar Lorentz-invariant theory of gravitation. *Archive for the History of the Exact Sciences* 45: no. 1: 17–94.

von Nostitz, Helene. 1966. *Potsdam.* Frankfurt: Weidlich. (See esp. pp. 61–62, "Die Sternwarten.")

Olbrich, Josef Maria. 1914. *Architektur, III. Serie.* Berlin: Wasmuth.

Onderdonk, Francis S. 1928. *The ferro-concrete style: Reinforced concrete in modern architecture.* New York: Architectural Books. (See esp. pp. 239–44 on contemporary tower buildings.)

Oppenheim, S. 1923. Karl Schwarzschild, zur 50. Wiederkehr seines Geburtstages. *VAG* 58: 191–209.

von der Pahlen, Emmanuel. 1924. Der unendliche Weltraum und die Relativitätstheorie. *Sterne* 4: 1–23.

———. 1926a. Der Cölostat des Einsteinturmes in Potsdam. *ZfI* 46: 49–67.

———. 1926b. Das Potsdamer Turm-Teleskop. *Himmelswelt* 36: 170–75.

von der Pahlen, E., and E. F. Freundlich. 1928. *Versuch einer dynamischen Bestimmung des K-Effektes sowie der Bewegungen im lokalen Sternsystem.* Publikationen des Astrophysikalischen Observatoriums, ser. 26, part 3 (no. 86).

———. 1930. Bemerkung zu einem Aufsatz von P. ten Bruggencate "The radial velocities of globular clusters." *ZAP* 1: 200–208. (Comment on ten Bruggencate 1930.)

Pais, Abraham. 1982. *Subtle is the Lord—The science and the life of Albert Einstein.* Oxford: Oxford University Press.

Parchomenko, P. 1926. Über das Strahlungsgleichgewicht der oberen Schichten der Sonne. *AN* 227: cols. 305–16. (See also reply in Freundlich et al. 1927.)

Paul, Erich Robert. 1993. *The Milky Way galaxy and statistical cosmology, 1890–1924.* Cambridge: Cambridge University Press.

Pehnt, Wolfgang. 1973. *Expressionist architecture.* London: Thames and Hudson.

———. 1981. *Architektur des Expressionismus.* 2d ed. Stuttgart: Hatje. (See esp. pp. 117–26 on Mendelsohn.)

Perrine, C. D. 1923. Contribution to the history of attempts to test the theory of relativity by means of astronomical observations. *AN* 219: cols. 281–84.

Peters, Gerhard. 1926. Die neue Baukunst in Deutschland. *Deutsche Monatshefte* 3: 161–73. (See esp. p. 168 on the Einstein Tower.)

Pettit, Edison. 1923. Focal changes in mirrors. *APJ.* 58: 208–14. (Contributions from the Mt. Wilson Solar Observatory, no. 266; on Pyrex glass.)

Platz, Gustav Adolf. 1927. *Die Baukunst der neuesten Zeit.* Berlin: Propylaen. (Propylaen-Kunstgeschichte, supplementary volume; see esp. p. 70 on the Einstein Tower.)

Plaut, Paul. 1929. *Die Psychologie der produktiven Persönlichkeit.* Stuttgart: Enke. (See esp. pp. 285–86 on Ludendorff and pp. 318–19 on Mendelsohn.)

Prager, Richard. 1926. Die Einrichtungen und Arbeiten der Sternwarte Berlin-Babelsberg. *Himmelswelt* 36: 140–53.

Price, Derek J. de Solla. 1963. *Little science, big science.* New York: Columbia University Press.

Pyenson, Lewis. 1974. The Goettingen reception of Einstein's general theory of relativity. Ph.D. dissertation, The Johns Hopkins University.

———. 1985. *The young Einstein: The advent of relativity.* Bristol: Hilger. (See also review in Hentschel 1985.)

Reitstätter, J. 1954. Freundlich, Herbert. *Kolloid-Zeitschrift* 139: 1–11.

Richter, Steffen. 1971. Das Wirken der Notgemeinschaft der deutschen Wissenschaft, erläutert am Beispiel der Relativitätstheorie in Deutschland 1920–1930. *PB* 27: 497–504.

———. 1972. *Forschungsförderung in Deutschland, 1920–1936, dargestellt am Beispiel der Notgemeinschaft der Deutschen Wissenschaft und ihrem Wirken für das Fach Physik.* Technikgeschichte in Einzeldarstellungen, no. 23.

———. 1979. Wirtschaft und Forschung: Ein historischer Überblick über die Förderung der Forschung durch die Wirtschaft in Deutschland. *Technikgeschichte* 46: 20–44.

Rimmer, W. B. 1932. Annual report of the Commonwealth Solar Observatory, Canberra, Australia. *MNRAS* 92: 295–97.

Ritschl, Rudolf. 1955. Walter Grotrian als Spektroskopiker. *Wiss.Ann.* 4: 561–71.

Robertson, Manning. 1925. *Laymen and the new architecture.* London: Murray. (See esp. pp. 74–75 on the Einstein Tower.)

Ross, Frank E. 1920. Image contraction and distortion on photographic plates. *APJ.* 52: 98–109.

Rowe, David E. 1989. Klein, Hilbert, and the Göttingen mathematical tradition. *Osiris* 5: 186–213.

Saal. 1901. Das Kuppelgebäude für den großen Refractor des astrophysicalischen Observatoriums auf dem Telegraphenberge bei Potsdam. *Zeitschrift für Bauwesen* 51: cols. 359–80 and figs. 40–42.

St. John, Charles Edward. 1917. The principle of generalized relativity and the displacement of Fraunhofer lines in the red. *APJ.* 46: 249–69.

———. 1922. Bemerkung zur Rotverschiebung. *Phys.Z.* 23: 197.

———. 1924. Zur Gravitationsverschiebung im Sonnenspektrum. *Z.Phys.* 21: 159–62.

———. 1928. Evidence for the gravitational displacement of lines in the solar spectrum predicted by Einstein's theory. *APJ.* 67: 195–239.

———. 1929. Commission 12. physique solaire. *Transactions of the International Astronomical Union* 3: 231–36.

Scheiner, Julius. 1890. Das königliche astrophysikalische Observatorium bei Potsdam. In *Die königlichen Observatorien für Astrophysik, Meteorologie und Geodäsie bei Potsdam*: 1–36, with pls. II–V. Berlin: Mayer and Müller.

Schlicker, Wolfgang. 1979. Konzeptionen und Aktionen bürgerlicher deutscher Wissenschaftspolitik: Zum gesellschaftlichen Stellenwert der Forschung nach 1918 und zur Gründung der Notgemeinschaft der Deutschen Wissenschaft. *Zeitschrift für Geschichtswissenschaft* 27: 423ff.

Schmidt, Paul Ferdinand. 1930. Erich Mendelsohn. *Cicerone* (Leipzig) 22: 219–25.

Schmitz, E.-H. 1984. *Handbuch zur Geschichte der Optik*. Vol. 4, part A. Bonn: Wayenborgh. (See esp. pp. 291–95 on tower telescopes.)

Schoenberg, Erich. 1948. Hans Ludendorff †. *Jahrbuch der Bayerischen Akademie der Wissenschaften*, 1944/48: 236–39.

Scholz, G. 1975. 100 Jahre astrophysikalisches Observatorium Potsdam—Arbeiten zur Spektroskopie. *Sterne* 51: 207–18. (See also Gußmann 1975, Wempe 1975.)

Schomerus, Friedrich. 1952. *Geschichte des Jenaer Zeisswerkes, 1846–1946*. Stuttgart: Piscator. (See esp. p. 138 on the Einstein Tower.)

Schroeder-Gudehus, Brigitte. 1978. *Les scientifiques et la paix: La communauté scientifique internationale au cours des années 20*. Montreal: Presses de l'Université de Montréal.

Schüler, Hermann. 1926. Über elektrische Anregung von Metalldämpfen im Kingschen Widerstandsofen. *Z.Phys.* 37: 728–31.

Schwarzschild, Karl. 1910. Die großen Sternwarten der Vereinigten Staaten. *Internationale Wochenschrift für Wissenschaft, Kunst und Technik* 4: cols. 1531–44.

———. 1914. Über die Verschiebung der Banden bei 3883 Å im Sonnenspektrum. *SB.Berlin*: 1201–13.

———. 1916a. Über das Gravitationsfeld eines Massenpunktes nach der Einsteinschen Theorie. *SB.Berlin*: 189–96.

———. 1916b. Über das Gravitationsfeld einer Kugel aus inkompressibler Flüssigkeit nach der Einsteinschen Theorie. *SB.Berlin*: 424–34.

von Seeliger, Hugo. 1891a. Notiz über die Strahlenbrechung der Atmosphäre. *SB.München* 21: 239–46.

———. 1891b. Ueber die Extinction des Lichtes in der Atmosphäre. Ibid.: 247–72.

———. 1893. Bemerkung über Strahlenbrechungen. *AN* 133: cols. 311–16.

———. 1906a. Über die sogenannte absolute Bewegung. *SB.München* 36: 85–137.

———. 1906b. Das Zodiakallicht und die empirischen Glieder in der Bewegung der innern Planeten. Ibid.: 595–622. (See also conclusion [ibid.: 594], Freundlich 1915a, and reply in von Seeliger 1915.)

———. 1913. Bemerkungen über die sogennante absolute Bewegung, Raum und Zeit. *VAG* 48: 195–201.

———. 1915. Über die Anomalien in der Bewegung der innern Planeten. *AN* 201: cols. 273–80. (Reply to Freundlich 1915a.)

———. 1916. Über die Gravitationswirkung auf die Spektrallinien. *AN* 202: cols. 83–86. (See also Ludendorff 1916.)

———. 1920/21. Fortschritte der Astronomie. *SDM* 18: 2–7.

Sharp, Dennis. 1966. *Modern architecture and expressionism*. London: Longmans, Green. (See esp. pp. 109–23 on the Einstein Tower.)

Siegmund-Schultze, Reinhard. 1984. Theodor Vahlen—Zum Schuldanteil eines deutschen Mathematikers am faschistischen Mißbrauch der Wissenschaft. *NTM* 21: 17–32.

Sigurdsson, Skuli. 1991. Hermann Weyl: Mathematics and Physics, 1900–1927. Ph.D. dissertation, Harvard University.

de Sitter, Willem. 1913a. Ein astronomischer Beweis für die Konstanz der Lichtgeschwindigkeit. *Phys.Z.* 14: 429.

———. 1913b. Über die Genauigkeit, innerhalb welcher die Unabhängigkeit der Lichtgeschwindigkeit von der Bewegung der Quellen behauptet werden kann. *Phys.Z.* 14: 1267.

Sommer, Richard. 1919. Astronomische Bestätigungen der Einsteinschen Relativitätstheorie. *Weltall* 19: 165–67. (Report on a lecture by Freundlich.)

Sommerfeld, Arnold. 1916. Karl Schwarzschild. *Natw.* 4: 453–57.

———, ed. 1923. *Das Relativitätsprinzip: Eine Sammlung von Abhandlungen*. 5th ed. Stuttgart: Teubner. (7th ed. 1974; contributions by H. A. Lorentz, A. Einstein, and H. Minkowski; supplement by H. Weyl; foreword by O. Blumenthal.)

Sontheimer, Kurt. 1968. *Antidemokratisches Denken in der Weimarer Republik*. Rev. ed. Munich: Nymphenburger.

Spieker, Paul. 1879. Die Bauausführung des königlichen Astrophysikalischen Observatoriums auf dem Telegraphenberge bei Potsdam. *Zeitschrift für Bauwesen* 29: 33–48.

Staal, Jan Frederick. 1920. Naar anleiding van Erich Mendelsohn's ontwerpen. *Wendingen* 3, no. 10: 3. (See also Beyer 1920.)

Stachel, John. 1986. Eddington and Einstein. In E. Ullmann-Margalit, ed., *The prism of science*: 225–50. Dordrecht: Reidel.

——— et al., eds. [J. Stachel, D. C. Cassidy, J. Renn, R. Schulmann, D. Howard, and A. J. Kox.] 1989a. *The collected papers of Albert Einstein*, vol. 2: *The Swiss years: Writings, 1900–1909*. Princeton: Princeton University Press (**b**) English trans. by A. Beck, ibid.

Staude, Jürgen. 1986. Spektroskopische Untersuchungen am Einsteinturm: Zur Physik der Korona und von Sonnenflecken. *Sterne* 62: 109–16.

———. 1991a. Das Sonnenobservatorium Einsteinturm in Potsdam. *Sterne und Weltraum* 67: 505–9.

———. 1991b. Solar Research at Potsdam: Papers on the structure and dynamics of sunspots. *Reviews of Modern Astronomy* 4: 69–89.

———. 1995. Sonnenforschung am Einsteinturm. *Sterne* 71: 142–50.

Stone, Sasha. 1990. *Fotografien, 1925–1939*. Ed. E. Köhn. Berlin: Nishen. (See esp. pp. 38–41 and 103–4 on the Einstein Tower.)

Stoney, G. Johnstone. 1896. On the equipment of the astrophysical observatory of the future. *MNRAS* 56: 452–59.

Strömgren, Elis. 1942. Hans Ludendorff †. *Natw.* 30: 53–55.

Struve, Hermann. 1919. *Die neue Berliner Sternwarte in Babelsberg*. Berlin. (Veröffentlichungen der Universitätssternwarte zu Berlin-Babelsberg, vol. 8, no. 1.)

———. 1975. Wahlvorschlag für Karl Schwarzschild. In Kirsten and Körber, eds., 1975, vol. 1: 195–96. (Undated.)

Tobies, Renate. 1981. *Felix Klein*. Leipzig: Teubner. (Biographien Hervorragender Naturwissenschaftler, Techniker und Mediziner, 50.)

———. 1994. Albert Einstein und Felix Klein. *Naturwissenschaftliche Rundschau* 47: 345–52.

Treder, Hans-Jürgen. 1974. Karl Schwarzschild und die Wechselbeziehungen zwischen Astronomie und Physik. *Sterne* 50: 13–19.

Trischler, Helmuth. 1992. *Luft- und Raumfahrtforschung in Deutschland, 1900–1970: Politische Geschichte einer Wissenschaft*. Frankfurt: Campus.

Trümpler, Robert J. 1932a. The deflection of light in the sun's gravitational field. *Publications of the Astronomical Society of the Pacific* 44: 167–73.

———. 1932b. Die Ablenkung des Lichtes im Schwerefeld der Sonne. *ZAP* 4: 208–20. (Comment on Freundlich et al. 1931; see also their response, 1932a.)

Trümpler, R., and Harold F. Weaver. 1953. *Statistical astronomy*. Berkeley: University of California Press.

Unsöld, Albrecht. 1928. Über die Struktur der Fraunhoferschen Linien und die quantitative Spektralanalyse der Sonnenatmosphäre. *Z. Phys.* 46: 765–81.

———. 1961. P. ten Bruggencate. *Jahrbuch der Göttinger Akademie der Wissenschaften*: 57–63.

———. 1972. *Sterne und Menschen: Aufsätze und Vorträge*. Berlin: Springer.

Vanýsek, V. 1976. Karl Wurm. *MAG* 38: 14–16.

Vierhaus, Rudolf, and Bernhard vom Brocke, eds. 1990. *Forschung im Spannungsfeld von Politik und Gesellschaft: Geschichte und Struktur der Kaiser-Wilhelm / Max-Planck-Gesellschaft*. Stuttgart: DVA.

Villiger, Walter. 1926. Das lichtstärkste Turmteleskop der Welt. *Zeiss Werkzeitung*, n.s., 2, no. 1: 5–6.

Virilio, Paul. 1992. *Das irreale Monument: Der Einstein-Turm*. Berlin: Merwe.

Vischer, Julius, and Ludwig Hilbesheimer. 1928. *Beton als Gestalter*. Stuttgart: Hoffmann. (See esp. pp. 17–18 on the Einstein Tower.)

Vogel, Hermann Carl. 1900. Description of the spectrographs for the great refractor at Potsdam. *APJ*. 11: 393–99, with pl. 9.

Vogt, H. 1920. Masse- und Dichteverhältnisse bei Doppelsternveränderlichen. *AN* 211: cols. 123–28.

Wali, Kameshwar C. 1991. *Chandra: A Biography of S. Chandrasekhar*. Chicago: University of Chicago Press. (See esp. pp. 94 and 122 on Freundlich.)

Walker, Mark. 1995. *Nazi science: Myth, truth, and the German atomic bomb*. New York: Plenum.

Weber, J. 1941. Prof. Dr. H. Ludendorff †. *Himmelswelt* 51: 126. (On Ludendorff see also ibid. 52 [1942]: 56–57.)

Weidle, Karl. 1929. *Goethehaus und Einsteinturm: Zwei Pole heutiger Baukunst*. Stuttgart: Zangg.

Wempe, Johann. 1975. Zum 100. Jahrestag der Gründung des Astrophysikalischen Observatoriums Potsdam. *Sterne* 51: 193–206. (See also Gußmann 1975, Scholz 1975.)

Westheim, Paul. 1923. Mendelsohn. *Kunstblatt* 4: 305–9.

Whittick; Arnold. 1940/56. *Eric Mendelsohn*. London: Hill. (**a**) 1st ed. (1940); (**b**) 2d ed. (1956; esp. pp. 54–58 on the Einstein Tower.)

Widmann, Horst. 1973. *Exil und Bildungshilfe: Die deutschsprachige akademische Emigration in die Türkei nach 1933*. Frankfurt: Lang.

Will, Clifford M. 1981. *Theory and experiment in gravitational physics*. Cambridge: Cambridge University Press.

——. 1986. *Was Einstein right? Putting general relativity to the test*. New York: Basic Books. (Reprinted Oxford: Oxford University Press, 1988. 2d ed., New York: Basic Books, 1993.)

Wolf, K. Lothar. 1927. Über eine Glühkathoden-Vakuumentladung in Gasen und Metalldämpfen. *Z.Phys.* 44: 170–89.

Wolfschmidt, Gudrun. 1991/92. Die Anwendung des Dopplereffektes in der Astronomie unter besonderer Berücksichtigung der Pionierleistung von H. C. Vogel. *NTM* 28: 173–209.

——. 1992/93. Kiepenheuers Gründung von Sonnenobservatorien im Dritten Reich. *Deutsches Museum, Wissenschaftliches Jahrbuch*: 283–318.

——. 1994. Sonnenphysik im zweiten Weltkrieg: Wissenschaft oder Kriegsforschung? In Christoph Meinel and Peter Voswinckel, eds., *Medizin, Naturwissenschaft, Technik und Nationalsozialismus: Kontinuitäten und Diskontinuitäten*: 152–59. Stuttgart: GNT.

Wright, Helen. 1972. Hale, George Ellery. *DSB* 6: 26–34.

Zevi, Bruno, ed. 1963. *L'Architettura—Cronache e Storia* 95 (anno 9, no. 5, Sept.: special issue on Erich Mendelsohn.)

———. 1985. *Eric Mendelsohn.* (**a**) London: Architectural Press; (**b**) New York: Rizzoli.

Index

In this index an "f" after a number indicates a separate reference on the next page, and an "ff" indicates separate references on the next two pages. A continuous discussion over two or more pages is indicated by a span of page numbers, e.g., "57–59." *Passim* is used for a cluster of references in close but not consecutive sequence. Italicized page numbers under name entries denote pages containing birth and death dates. References to Albert Einstein, the Einstein Tower, and Erwin F. Freundlich are not included in this index.